Ethics and the Good Soldier

Building on extensive, internationally leading empirical research conducted by the Jubilee Centre for Character and Virtues, this book explores the soldier as a virtuous professional through a close examination of soldiers' character and ethical reasoning.

Starting from the view that virtues such as courage, honesty, loyalty and integrity are core to being a professional soldier, the book draws on insights from British soldiers at three stages of their career – officer cadets (at Sandhurst), junior lieutenants/captains (attending the Junior Officer Tactical Awareness Course) and senior captains (participating in the Captain's Warfare Course). Drawing upon the Jubilee Centre's *Soldiers of Character* study, the book explores soldiers' character in the professional domain.

Including clear implications for practice and research, this is essential reading for all those interested in the character of soldiers in both a professional and an academic setting.

James Arthur is Professor of Education at the University of Birmingham.

Scott Parsons is Assistant Vice Chancellor for Character and Ethics Development at the Texas Tech University System. He is a retired US Army Intelligence Officer and spent 2014–2023 as Assistant Professor and Character Development Integrator at the United States Military Academy at West Point.

Andrew Peterson is Professor of Character and Citizenship Education in the Jubilee Centre for Character and Virtues at the University of Birmingham, UK.

Character and Virtue within the Professions

Series Editors

James Arthur
Professor of Education at the University of Birmingham.

Andrew Peterson
Professor of Character and Citizenship Education and Deputy Director of the Jubilee Centre for Character and Virtues at the University of Birmingham, UK.

The principal objective of the series is to highlight the interplay between practitioners' personal character and the ethical dimensions of their professional domain. Each book will explore the specific ethical dimensions of the given profession at hand, including the interplay between professionals' individual character virtues and their working environments. In a time when cultures of managerialism, auditing, performance metrics and commercial success are seemingly increasing, this series attempts to re-focus the professions towards the ethical and societal origins that each profession intends to serve. Underpinned by perspectives of philosophy, psychology and sociology, each book will offer practitioners fresh viewpoints about how their character and professional context can influence their professional practice.

Books in the series include:

Ethics and the Good Teacher
Character in the Professional Domain
Andrew Peterson with James Arthur

Ethics and the Good Doctor
Character in the Professional Domain
Dr. Sabena Jameel, Andrew Peterson and James Arthur

Ethics and the Good Nurse
Character in the Professional Domain
Andrew Peterson, James Arthur and Jinu Varghese

Ethics and the Good Soldier
Character in the Professional Domain
James Arthur, Scott Parsons, and Andrew Peterson

For more information about this series, please visit: www.routledge.com/Character-and-Virtue-Within-the-Professions/book-series/CVP

Ethics and the Good Soldier

Character in the Professional Domain

James Arthur, Scott Parsons, and Andrew Peterson

LONDON AND NEW YORK

First published 2024
by Routledge
4 Park Square, Milton Park, Abingdon, Oxon OX14 4RN

and by Routledge
605 Third Avenue, New York, NY 10158

Routledge is an imprint of the Taylor & Francis Group, an informa business

British Library Cataloguing-in-Publication Data
A catalogue record for this book is available from the British Library

ISBN: 978-1-032-36298-4 (hbk)
ISBN: 978-1-032-36299-1 (pbk)
ISBN: 978-1-003-33120-9 (ebk)

DOI: 10.4324/9781003331209

Typeset in Times New Roman
by Apex CoVantage, LLC

Contents

Charts

Figures

Tables

Acknowledgements

This book is the fourth in a series of texts that examine *Character and Virtues in the Professions*. Each book in the series is dedicated to a specific profession and brings together reviews of existing literature and sources of empirical data collected by the Jubilee Centre for Character and Virtues to provide new insights for both pre- and in-service professionals, as well as acting as an educational resource to inform future professional decision-making and practice.

As we make clear from the outset, this book draws extensively on data and analysis from one of a number of research projects on virtues in the professions conducted and reported on by the Jubilee Centre that were all led by the Centre's Director, James Arthur. For this reason, we owe our sincere gratitude to those colleagues whose data collection, analysis, recommendations and overall insight on the project reported here, including in the resulting report, have made this book possible. In particular, our thanks and acknowledgements go to David Walker and Steve Thomas. We are also grateful to our colleagues at Routledge – in particular, Anna Clarkson, Sarah Hyde and Rhea Gupta – for their interest and support in this series and book.

James Arthur
Scott Parsons
Andrew Peterson

Introduction

Introduction

When one thinks of examples of a profession, several come quickly to mind: doctors, nurses, teachers, clergy, lawyers, business people and members of the police force. Yet, standing armies have existed for well over 2,000 years, and today national armies typically refer to their members as professionals. While it is commonplace to regard soldiers as members of a profession, defining and delineating the precise ethical contours of the profession remain contentious. In his classic book, and as we draw on within the chapters of this book, General Sir John Hackett (1983: 9) referred to soldiers as members of the 'profession of arms', arguing that 'the function of the profession of arms is the ordered application of force in the resolution of a social or political problem', while in his book *The Soldier and the State*, Samuel P. Huntington (1957: 11) defined officers, and all soldiers for that matter, as professionals, contending that 'the modern officer corps is a professional body and the modern military officer is a professional'. Similarly, Morris Janowitz (1960: 5) positioned soldiers as 'professionals in violence'. More recently, a now quite extensive literature exists that is interested in the army as a profession that has developed across a range of contexts (see, for example, Burk, 2002; Challans, 1999; Collins; 2007; Huntington, 1963; Miller, 2004; Moten, 2011; Robinson, 2008; Snider et al., 1999; Snider and Watkins, 2000, 2002; Swain and Pierce, 2019; Wilson and Meese, 2019).

This book, and the main study from which the data contained within it is drawn, starts from the premise that being a soldier is a complex moral undertaking that raises significant questions about conduct and – crucially so far as this book is concerned – character. In other words, being a soldier involves much more than following certain rules of engagement but necessarily brings into focus specific virtues, including courage, loyalty, honest and service. Certainly, being a member of a professional army can place soldiers in a multitude of situations in which ethical responses are not only required but also place serious burden and challenge on soldiers. As we will suggest in this book, and as others have argued (see, for example, Olsthoorn, 2011;

DOI: 10.4324/9781003331209-1

Walker, 2020, 2021; Walker et al., 2021), in such morally complex situations (that are perhaps further removed from the experiences of non-members than any other profession), rules and codes of conduct can only carry so much of the load required. On this reading, being a 'good' soldier goes beyond tactical and operational excellence and requires various qualities of character – or virtues – that enable soldiers to respond well to the situations faced.

These virtues lie at the heart of armies and/or armed forces more generally, and while not completely unrelated to ethical codes of conduct, they suggest that specifying what it means to be a "good" soldier in an ethical sense goes beyond following pre-established rules. Indeed, armies typically prioritise a particular set of dispositions or values viewed as central to the conduct of soldiers (as depicted in Table 0.1).

Not only do modern armed forces list values soldiers are to live and conduct themselves by, the armed forces in Table 0.1 also offer definitional clarification about what the values mean and how they are to be understood, typically through the provision of short definitions – which themselves often include references to character. The Australian Army, for instance, defines its five core values as follows:

- **Service:** The selflessness of character to place the security and interests of our nation and its people ahead of my own.
- **Courage:** The strength of character to say and do the right thing, always, especially in the face of adversity.
- **Respect:** The humanity of character to value others and treat them with dignity.
- **Integrity:** The consistency of character to align my thoughts, words and actions to do what is right.
- **Excellence:** The willingness of character to strive each day to be the best I can be, both professionally and personally.[8]

Similarly, the British Army offers the following short definitions of their six values:

- **Courage:** Doing and saying the right thing not the easy thing.
- **Discipline:** Doing things properly and setting the right example.
- **Respect for Others:** Treat others as you expect to be treated.
- **Integrity:** Being honest with yourself and your teammates.
- **Loyalty:** Support the army and your teammates.
- **Selfless commitment:** Mates and mission first, me second.[9]

In addition to these short definitions, the more substantial definitions of each of the values of the British Army further demonstrate the centrality of character within the values. The value of courage, for instance, is clearly specified as both a physical and moral attribute. Moral courage is defined as 'doing

Table 0.1 Values of Different National Armies

Australian Army's Core Values[1]	British Army's Values[2]	Canadian Forces' Essential Values[3]	New Zealand Army' Values[4]	Norwegian Armed Forces' Values[5]	Spanish Land Army's Values[6]	US Army's Values[7]
Service	Courage	Duty	Courage	Respect	Valour	Loyalty
Courage	Discipline	Loyalty	Commitment	Responsibility	Spirit of Sacrifice	Duty
Respect	Respect for Others	Integrity	Comradeship	Courage	Discipline	Respect
Integrity	Integrity	Courage	Integrity		Companionship	Selfless Service
Excellence	Loyalty				Spirit of Service	Honor
	Selfless Commitment				Honour	Integrity
						Personal Courage

the right thing, not looking the other way when you know or see something wrong, even if it is not a popular thing to do or say'.[10]

Considerations of character and virtues matter in the day-to-day lives and experiences of soldiers and militaries. At times, and often as a result of high-profile incidents and resultant scandals (whether to do, for example, with the unlawful treatment of civilians and prisoners of war, sexual harassment and assaults within the army, bullying or the (mis)use of social media), the morality and conduct of soldiers receive wider public attention. Also of importance in clarifying the ethical dimensions of the good soldier is the recognition that various changes in modern warfare raise different (or at least differentiated) challenges for professional armies today. These changes are multifarious and include the changing composition of modern armies (e.g. the ending of conscription and compulsory national service) and the ways that technology has impacted on combat and military engagements. More detailed consideration to these changes is given in Chapter 2.

Since its inception in 2012, the Jubilee Centre for Character and Virtues at the University of Birmingham has conducted numerous studies that have sought the views, perceptions and explanations of professionals themselves in order to interrogate and explore the ethical dimensions of professions in England, each led by the Centre's Director, Professor James Arthur. In this book, we draw on data from one of these studies – *Soldiers of Character* – to present, analyse and explore how soldiers conceive the moral dimensions of their work. While we add some additional analysis to the findings, including reporting additional qualitative data from the study and entering into conversation with relevant research literature that has become available since the *Soldiers of Character* report was published, this book draws extensively from the original research report (Arthur et al., 2018).

In addition to the introductory text above, this chapter has two purposes. The first purpose is to set out, largely in summary given the available space, the core details of the main project on which the analysis offered in the book is based. For concision, the main aims, research questions and methods employed are specified. The detailed aims, research questions and methods of the project can be found on the Jubilee Centre's website,[11] as can more details of its extensive research on the professions more generally.[12] The second purpose is to detail the structure of the book, including the focus and broad content of each of the chapters that follow.

The Project

The *Soldiers of Character* project was an empirical study of over 240 junior British Army officers from 12 branches of service conducted between 2015 and 2017. The overarching aim of the *Soldiers of Character* project was to investigate how far junior officers in the British Army displayed and aspired to personal characteristics described in the Army Values and Standards

Guide. The study also examined a broader range of character strengths among all ranks from the perspective of junior officers. The main research questions that guided the project were as follows:

1. To what extent do junior officers show ethical reasoning in line with standards of excellence described in the British Army Values and Standards Guide, especially regarding strengths of courage, respect for others, integrity and loyalty?
2. How do junior officers rate their own character strengths? What are their most and least dominant strengths?
3. How do responses to moral dilemmas relate to the junior officers' self-reported character strengths and to questions asked during the interview about Army values?
4. How do high- and low-performing (Army Intermediate Concept Measure (AICM)) junior officers relate to values of selfless commitment and discipline? What routine and key professional challenges have these officers faced, and what lessons (if any) were learnt? What qualities of an 'ideal' junior officer are admired and aspired to? To what extent do officers believe that Army values transfer across professional and personal lives?

The project employed a combination of the following three methods, each of which is set out in more detail in this section: (1) moral dilemmas (AICM), (2) self-reports of character strengths (VIA-IS-E1) and (3) semi-structured interviews. All respondents involved in the study completed the first two of these, with a sub-sample also participating in the third.

Demographics

While 97 per cent of the sample involved in the study was white, the sample was diverse in terms of Regiment or Corps/Branch of Service, gender, rank and length of service (see Appendix 2). The distribution of gender and branch of service across rank and experience is shown in Table 0.2. Although the British Army divides branches of service into Combat Arms, Combat Support and Combat Service Support, for the purposes of this *Soldiers of Character* project, officers were divided into two groups: those who belong by cap badge to artillery or infantry regiments and those who do not, on the basis that this distinction was associated with the most noticeable differences in AICM results.

Moral Dilemmas: AICM

Moral dilemmas have been employed as a research tool in various Jubilee Centre studies on the professions (see, for example, Arthur et al., 2014, 2015; Kristjánsson et al., 2017a, 2017b), as well as in studies by others (see, for instance, Bebeau and Thoma, 1999). The use of moral dilemmas in the

Table 0.2 Gender and Branch of Service by Rank/Experience

		Cadet	Lieutenant and Junior Captain	Senior Captain and Major	Total
Gender by rank and experience	Male	57 (75%)	71 (76%)	62 (84%)	190 (78%)
	Female	19 (25%)	22 (23%)	11 (15%)	52 (21%)
	Total	76	93	73	242
Branch of service by rank/ experience	Infantry/ artillery	23 (45%)	32 (35%)	32 (44%)	87 (41%)
	Non-infantry/ artillery	28 (55%)	59 (65%)	41 (56%)	128 (60%)
	Total	51[13]	91	73	215

Note: Percentage within rank is shown in brackets.

Soldiers of Character study employed an intermediate concept approach, as developed by Rest et al. (1999a and 1999b). As explained in more detail in Chapter 3, intermediate concepts are situated between 'bedrock' schemas of moral reasoning (self-interests, maintaining norms and post-conventional schemas) and specific contextual norms (such as professional codes). More specifically, the AICM that was administered sought to bridge neo-Kohlbergian (a psychological approach to moral development) and neo-Aristotelian (a traditional philosophical theory of moral and character development) approaches to moral reasoning and justifications by asking respondents to make moral judgements about a story in which a virtue (Army value) is at stake. Following other studies, the Intermediate Concept Measure (ICM) dilemma tests rest on the idea that 'patterns of ratings and rankings in response to the dilemmas reveal information about the extent to which participants' application of virtue concepts match expert views' (Arthur et al., 2018: 15) rather than seeking to assess moral schemas directly.

The study drew on the Army Leadership Ethical Reasoning Test (ALERT) Army ICM, developed first at the University of Alabama for junior US Army officers serving at the United States Military Academy West Point Military by Lieutenant Colonel Michael Turner, and which involved a panel of senior experts in ethical judgement in US military contexts to create dilemmas and associated items. For the *Soldiers of Character* study, the project team moved to four, rather than seven, dilemmas, with each of the dilemmas revised for applicability to the context of the British Army. The four dilemmas (which can be found in Appendix 1 and are analysed in Chapter 3) focused on dominant British Army values, and each was accompanied by a given set of action choices *and* reasons on a scale from 1 (*I strongly believe that this is a GOOD choice/reason*) to 5 (*I strongly believe that this is a BAD choice/reason*) (again, which can be found in Appendix 1). Officers then selected and ranked *best/most important* (first, second and third) and *worst/least important* (first, second and third) options for actions *and* reasons. Demographic questions were also asked (details of the approach to validating the AICM measure can be found in the project's final report (Arthur et al., 2018)).

The process of determining the final dilemmas to be included and their substance involved the following phases:

1. Consultation with British Army experts in ethics, psychology and law
2. The bringing together of an expert panel in a British Army garrison. The panel comprised 11 lieutenants and captains with different lengths of experience. The expert panel considered and revised the dilemmas to be applicable to the British Army in July 2015.
3. A further expert panel was constituted at another British Army garrison. This second expert panel comprised 12 lieutenants and captains, again with different lengths of experience. This panel assessed the dilemmas as amended in phases 1 and 2, but after also completing the whole survey individually. This took place in September 2015.
4. The research team compared all three expert panels (the two UK panels plus the original US expert panel for ALERT) in order to ensure that the four dilemmas were credible, realistic and believable for British Army officers. This phase also involved the development of a key based on agreement across panels.

Junior Army officers completed the AICM, supervised by members of the research team. The completed AICMs were subjected to basic automated analysis to produce results on the basis of scores of 'acceptable', 'neutral' or 'unacceptable' as determined by the expert panel process. For instance, and as explained in the final report of the project, always

> selecting 'acceptable' options as good and 'unacceptable' options as bad produced a score fully compatible with the expert panel (100%); selecting appropriate choices in this way for half of the required choices will produce a score of about 50%; and selecting 'neutral' options will not raise or lower the score.
>
> (Arthur et al., 2018: 16)

Self-Reporting Measure – VIA-IS-E1

The project gathered self-report data using the Values in Action (VIA) measure (Peterson and Seligman, 2004), employing the VIA-IS-E1 tool. The study used this method to determine an understanding of the officers' character strengths. The VIA-IS-E1 consists of 24 questions, and the officers respond to these after the AICM and within the same survey. For this, the officers responded to the statement: 'This strength is an essential part of who I am in the world' in relation to 24 specific character strengths (e.g. perseverance) by choosing from the given options (*strongly agree* (5), *definitely agree* (4), *slightly agree* (3), *neutral* (2), *disagree* (1)). In the analysis undertaken by the research team, and presented in this book, the overall results were averaged to determine how certain groups of officers differed in relation to the 24 character strengths.

Semi-structured Interviews

The in-depth, semi-structured interviews enabled the research team to examine in more detail and depth how officers understood character and virtues in the army. A core aspect of the interview design was to engage officers in giving examples from their own experiences, including how they understood and observed virtues and conduct. A sub-sample of officers undertaking the survey was interviewed, comprising a mix of all three officer experience levels based on a purposive sampling approach. The interviews were audio-recorded with participant consent and later transcribed. For the purpose of the project's final report, thematic analysis was carried out for the top and bottom ten scoring officers based on their total AICM scores (Braun and Clarke, 2006). Data pertaining to this were coded using Nvivo software. For the purposes of this book, interview transcripts were revisited and subjected to further thematic analysis.

Recruitment and Access

Data were collected at three key Army courses in 2016 where representative participants clustered. These were Sandhurst for officer cadets, the Junior Officer Tactical Awareness Course (JOTAC) for junior lieutenants/captains and the Captain's Warfare Course (CWC) for senior captains. A small number of participants were also recruited from a UK Army garrison. As reported in the project's final report (Arthur et al., 2018), a stratified random sampling approach was used within the three levels of Army experience. The decision was taken that women were oversampled in order to assess the possibility of gender differences for AICM, on the basis that women have been found to consistently outperform their male peers in moral dilemma measures such as the ICM.

Limitations and Ethical Considerations

The first limitation to note is that while the AICM as a measure of ethical reasoning has good support prior to and following this research, the measure needs further testing before it may be considered fully validated. A second limitation is that the *Soldiers of Character* study sort to examine one level of the chain of command – junior officers. As pointed out in the original report (Arthur et al., 2018), it is likely to be the case that interactions with other levels of the chain of command will influence the character, virtues and values of officers. We return to this point in the final chapter of this book.

Ethical approval for the research reported here was granted by the Ministry of Defence Research Ethics Committee. Informed consent was also obtained from all participants. As once again set out in the project's final report, some specific ethical concerns unique to Army environments required careful handling in the conduct of the project. For instance, consideration had to be given to potentiality that participants might be suffering from PTSD or other mental health conditions which may have become apparent during the interviews. The project term constituted clear and careful plans for such situations, and these were agreed with the Army. Participants were guaranteed confidentiality and anonymity and could withdraw their involvement up to a given date.

The Structure of This Book

Following this introduction, the book consists of four main chapters and a conclusion. The conclusion serves to summarise the main findings, offers some recommendations on the basis of the study (including and extending those offered in the project's final report) and identifies some areas for possible further research. Chapter 1 provides a brief review of current literature in the wider field of professional ethics and in doing so sets out the core argument that ethics, and being ethical, is central to what it means to be a profession. The chapter also introduces the recent turn to virtues and character within literature on the professions, in particular the focus on *phronesis* (practical wisdom). In this context, the Jubilee Centre's *Building Blocks of Professional Practice* are also introduced. The focus in Chapter 2 moves to a more specific consideration of the military, and more specifically the army, as a profession. Drawing on current literature in the field, the chapter considers some central ideas within the field of military ethics, with a specific focus on the Army and on virtue-based understandings of the ethical soldier. The chapter examines the military as the 'profession of arms', general military ethics, virtue ethics and the military and the changing nature of soldiering and military ethics. The chapter takes a particular interest in how the conduct and character of soldiers have been positioned in the literature and also considers these in relation to various aspects of wider military ethics.

Chapters 3 and 4 present empirical data gathered through the *Soldiers of Character* project. Chapter 3 examines survey data from the AICM and VIA-IS-EI measures to evidence the soldiers' ethical reasoning about the four dilemmas, including how this reasoning stood in relation to the best and worst action and justification decisions posited by the expert panel. In addition, the chapter presents data drawing on soldiers' self-reports of virtues important to who they are using the VIA-IS-EI measure. Chapter 4 presents a narrative, thematic analysis of data obtained through semi-structured interviews with 40 officers as part of the *Soldiers of Character* study. The interviews examined how the officers perceived and experienced values, virtues and character in the British Army. In the chapter five key themes are explored: the officers' motivations for joining the Army/becoming an officer; their views on the extent of transfer of Army Values and Standards across professional and personal lives; the personal qualities and character strengths an ideal officer of their own rank might have; the personal qualities or strengths most important to them in their current role as an Army officer; and the pressures or barriers that make it difficult for them – or others like them – to act ethically. The book closes with 'Conclusions, Recommendations and Further Research'.

Our intention in the pages that follow is to bring together data, analysis, findings and conclusions from the *Soldiers of Character* project in order to consolidate and share with new audiences what the project revealed about the ethical dimensions of being a soldier and of the army as a profession.

Notes

1 www.army.gov.au/our-people/our-values-contract
2 www.army.mod.uk/who-we-are/our-people/a-soldiers-values-and-standards/
3 www.canada.ca/en/department-national-defence/services/benefits-military/defence-ethics/policies-publications/code-value-ethics.html
4 www.nzdf.mil.nz/army/
5 www.forsvaret.no/en/about-us/missions-and-values/values
6 https://ejercito.defensa.gob.es/en/personal/valores/index.html?__locale=en
7 www.army.mil/values/
8 www.army.gov.au/our-people/our-values-contract
9 www.army.mod.uk/who-we-are/our-people/a-soldiers-values-and-standards/
10 www.army.mod.uk/who-we-are/our-people/a-soldiers-values-and-standards/
11 www.jubileecentre.ac.uk/1582/projects/current-projects/soliders-of-character
12 www.jubileecentre.ac.uk/1595/projects/virtues-in-the-professions
13 Some officer cadets did not know to which branch of service they would be allocated.

1 The Professions and Character

Introduction

Whether inspired by a desire to justify an occupation's status as a profession (teaching and social work, for example) or by the need to re-assert precisely what lies at the heart of a long-standing profession in the wake of public concerns about standards (medicine and law, for example), the related questions of what constitutes a profession and what constitutes professional practice have received a great deal of attention over recent years. A core concern within this literature on the professions has been to highlight and seek to understand the ethical basis of professions, whether generally or specifically. Professions are deemed inherently ethical occupations because, and more so than other occupations, they place high moral demands on the conduct of workers. Indeed, these ethical and moral demands – which include care, integrity, fairness and diligence – are often viewed as *the defining* features of many professions, including nursing, medicine, law, teaching and the military, reminding us that professions are ultimately concerned with *human* actions and interactions. For example, in relation to medicine, Bontemps-Hommen et al. (2019) have suggested that morality is at the heart of medicine. As Oakley and Cocking (2001) also remind us, the focus of professional work is typically the provision of goods – such as health, education and justice – that are fundamental to flourishing individuals and societies. In specific relation to healthcare, a number of authors cite the importance of virtues such as trust, compassion and kindness as being core to the profession (see, for example, Armstrong, 2007; Brody and Doukas, 2014; Rhodes, 2020; Tuckett, 2000). Yet, as various professional "scandals" over the last 20 years have evidenced, every profession and professional faces ethical challenges and dilemmas as part of their work. Indeed, the very ethical nature of the professions entails that public mistrust and criticism result when conduct falls below the standards expected (Blond et al., 2015).

In order to examine the ethical nature of professions and the ethical dilemmas experienced by professionals, since its inception, the Jubilee Centre has

DOI: 10.4324/9781003331209-2

undertaken a number of empirical studies examining character, virtues and the professions. Some of these studies have concentrated on the professions generally (Arthur et al., 2020), while others have focused on specific professions: law (Arthur et al., 2014), medical practice (Arthur et al., 2015a), education (Arthur et al., 2015b), business (Kristjánsson et al., 2017b) and the British Army (Arthur et al., 2018). More recently, through the project *Practical Wisdom and Professional Practice: Integration and Intervention*, the Centre has built on this research to examine particular commonalities and differences across professions and professionals (Arthur et al., 2020).

The purpose of this first chapter is to provide an initial survey of the existing literature on the professions. The first section considers briefly what constitutes a profession in general terms before turning to the more specific ethical dimensions of professional activity. It does so in light of the now widespread trend towards managerialism, accountability and efficiency that has been witnessed across professions in a number of countries over the last 30 years. In the second section, attention moves to considering the value of a virtue-based account of professional ethics. In this section we draw on the Jubilee Centre's neo-Aristotelian approach to virtues and character in order to argue that professional ethics not only involves but also transcends reliance on rules and duties, thereby requiring professionals to act with professional wisdom and judgement.

While we engage with existing literature on the army as a profession in much more detail in Chapter 2, at various points in this current chapter we make initial and more general references in order to offer some initial starting points and to bring these ideas into relation with other professions.

What Constitutes a Profession?

While definitions of what constitutes a profession abound, certain features seem to be generally, if not universally, accepted (see, for example, Carr, 1999). These are:

- A profession is a paid occupation
- A profession requires formal qualifications, a high level of education and a prolonged period of training/induction
- A professional possesses high-level theoretical and practical expertise in a given discipline
- A profession provides a public service
- A profession is, and professionals are, held in high esteem within society
- A professional acts with integrity, care, honesty and trust, exhibiting a level of professional autonomy and judgement
- Professional ethics is guided by a code of conduct specific to that profession

The Australian Council of Professions,[1] which captures each of the features above, defines a 'Profession' as

> a disciplined group of individuals who adhere to ethical standards and who hold themselves out as, and are accepted by the public as possessing special knowledge and skills in a widely recognised body of learning derived from research, education and training at a high level, and who are prepared to apply this knowledge and exercise these skills in the interest of others. It is inherent in the definition of a Profession that a code of ethics governs the activities of each Profession. Such codes require behaviour and practice beyond the personal moral obligations of an individual. They define and demand high standards of behaviour in respect to the services provided to the public and in dealing with professional colleagues. Further, these codes are enforced by the Profession and are acknowledged and accepted by the community.

In the UK, various professions make clear the centrality of the 'ethical' to the nature of the profession. For example, in its Code of Ethics,[2] the British Association of Social Workers asserts:

> Ethical awareness is fundamental to the professional practice of social workers. Their ability and commitment to act ethically is an essential aspect of the quality of the service offered to those who engage with social workers. Respect for human rights and a commitment to promoting social justice are at the core of social work practice throughout the world.

The Law Society of England and Wales[3] makes clear:

> The commitment to behaving ethically is at the heart of what it means to be a solicitor.
> Ethics is based on the principles of:
>
> * serving the interests of consumers of legal services
> * acting in the interests of justice acting with integrity and honesty according to widely recognised moral principles
>
> Ethics will help you respond in the right way to any moral dilemmas you might face at work.

Turning more specifically to the Army, for example, the United States Army publishes doctrine through their Army Publishing Directorate. Currently, there are 13 Army Doctrine Publications (ADPs). Additionally, the US Army has Army Doctrine Reference Publications (ADRPs), which serve as a way to add specificity to the ADPs. ADRP 1, titled *The Army Profession*, starts with a quote by then Chief of Staff of the Army General Raymond T. Odierno, '[We will] foster continued commitment to the ***Army Profession***, a noble and selfless calling founded on the bedrock of trust' (emphasis is that of the authors). Sections 1–2 of ADRP 1 states that 'a profession is a trusted, disciplined, and relatively autonomous vocation whose members:

- Provide a unique and vital service to society, without which it could not flourish
- Provide this service by developing and applying expert knowledge
- Earn the trust of society through ethical, effective, and efficient practice
- Establish and uphold the discipline and standards of their art and science, including the responsibility for professional development and certification
- Are granted significant autonomy and discretion in the practice of their profession on behalf of society'

Additionally, Sections 1-10-1-16 of ADRP 1 state:

- The Army Profession is a unique vocation of experts certified in the ethical design, generation, support and application of landpower, serving under civilian authority and entrusted to defend the Constitution and the rights and interests of the American people.
- An Army professional is a Soldier or Army Civilian who meets the Army Profession's certification criteria in character, competence, and commitment.
- The Army, like other professions, inspires and motivates its members to make right decisions and take right action according to the moral principles of its ethic.

Other professions that similarly locate ethical conduct as fundamental to the profession could be cited. However, despite these reasonably well-established and understood definitions, how best the ethical should be formulated conceptually and can be implemented practically remains both disputed and challenging – whether in terms of individual professions or the professions more generally.

Clearly, ideas about what constitutes a 'good' professional transcend simply technical abilities and encompass notions of judgement, wisdom and care. However, further questions remain about the extent to which particular cultures, discourses and practices can put pressure on how professionals, particularly those working in the public sector, can act with (or indeed without) ethics and integrity (see, for example, Furlong et al., 2017). Indeed, various studies evidence the impact (whether positive or negative) of workplace conditions on professionals' ability to exhibit ethical conduct (see, for example, Oakley and Cocking, 2001; RPS, 2011; Worth and Van Den Brande, 2019).

Discussions about the meaning and nature of ethical professional conduct and the effect of cultures, discourses and workplace practices typically concentrate around two particular considerations. The first is the impact, widely cited and critiqued in current literature on ethics and the professions, of the increased forms of managerialism and instrumentalism that have roundly been identified as detracting from the ethical and societal role of professionals. According to critics, the turn to managerialism across and within the professions has led not to a renewed form of professionalism but to a processes of de- and re-professionalisation through which the goals of general accountability (to service-users and to government) and efficiency have actively worked against professional autonomy and judgement (Carr, 2011; Holbeche and Springett, 2004). The second consideration is the extent to which professions, such as health, teaching and social work, have come under increased public scrutiny and accountability in the wake of various 'scandals' (Seijts et al., 2017). Over the last 25 years in England, for example, high-profile cases including the murder of Stephen Lawrence and the resulting Stephen Lawrence Inquiry (known commonly as the Macpherson Report), the murder of Victoria Climbié, the death of Peter Connelly (also known as Baby P), the Mid Staffordshire hospital crisis and the Rotherham Child Sexual Exploitation scandal and controversies about misconduct in the Metropolitan Police have all raised serious questions about what were significant failures in professionals' ethical judgement and conduct.

In the context of managerialism, accountability, efficiency, public scrutiny and increased workplace pressures, professions and professionals need to (re)envisage the ethical nature of their work. This (re)envisaging by necessity includes paying attention to what a profession aspires to be, what constitutes professional practice – whether generally or specifically for that profession – and how external factors shape the standing and work of professions today. In the next section we start to examine these questions through a focus on a virtue-based approach to professional ethics. In doing so, we introduce key work in the field, particularly that which makes reference to the concept of professional *phronesis*.

A Virtue-Based Approach to Professional Ethics

The last few decades have witnessed a groundswell of interest in virtue-based approaches to professional ethics. Though not the only variant of a virtue-ethical approach, the vast majority of this interest has drawn on Aristotelian roots, and this concerted interest in Aristotelian/neo-Aristotelian virtue has been applied across a range of professional contexts that are not the immediate focus of this present book, including accountancy (e.g. West, 2017), medicine (e.g. Pellegrino and Thomasma, 1993), nursing (e.g. Armstrong, 2007), social work (e.g. Adams, 2009), and youth work (e.g. Bessant, 2009). In particular, two Aristotelian ideas have provoked significant interest among those concerned with professional ethics. The first is the idea that virtues represent 'contextually appropriate traits . . . such as honesty, compassion and perseverance' that contra rules 'become habitually ingrained through deliberate and repetitive practice, predisposing practitioners to behave based on ethically sound habits' (Arthur et al., 2019: 2). The second idea – the main focus of this section – is the concept of *phronesis*, or practical wisdom (Annas, 2011; Aristotle, 1985; Gillies, 2005; Kinsella and Pitman, 2012; McKie et al., 2012; Pellegrino and Thomasma, 1993). It is important to note, however, that while often cited, *phronesis* is not understood *uniformly* throughout the literature on professions. Indeed, examining work on *phronesis* in professional medical ethics, Kristjánsson (2015a: 299) highlights the 'considerable lack of clarity in the current discursive field on *phronesis*' (there has been some discussion of phronesis in relation to the Army; see, for instance, Zacher et al. (2015); DeFalco and Doty (2019); Arthur et al. (2021); Walker (2018, 2020, 2021); Parsons (2021)).

In line with its neo-Aristotelian philosophy, the Jubilee Centre advocates the following model of the **Building Blocks of Professional Practice** (see Figure 1.1).

The Jubilee Centre's *A Framework for Character Education in Schools* (2017) was adapted to a professional domain. The model depicts the four domains of virtue and their conceptual relationship with practical wisdom and purposeful professional practice.

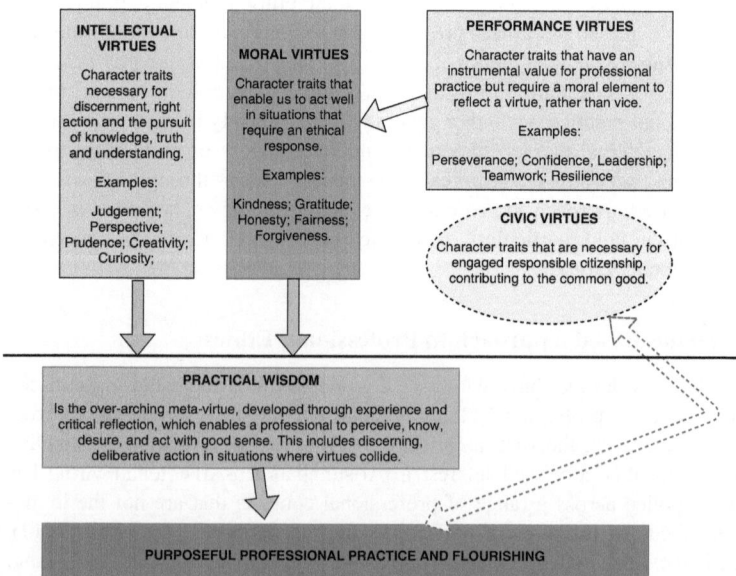

Figure 1.1 The Building Blocks of Professional Practice

In Figure 1.1, *phronesis* – or practical wisdom – is defined as 'the over-arching meta-virtue, developed through experience and critical reflection, which enables a professional to perceive, know, desire and act with good sense. This includes discerning, deliberative action in situations where virtues collide'. In other words, professionals need a certain form of practical wisdom, or *phronesis*, which can be defined in the following way:

> To practice with *phronesis* is to act with care, diligence and open-mindedness.
>
> To practice without *phronesis* would mean acting carelessly, indecisively, and with a degree of negligence to the surrounding circumstances or possible consequences.
>
> (Arthur et al., 2019: 5)

For some authors, it is possible and useful to identify a form of professional *phronesis* – or what Sellman (2009: 1) terms the 'professionally wise practitioner'. Sellman (2012: 116) defines the professionally wise practitioner as one who

> continually strives to be the best practitioner she or he can be given the constraints under which practice occurs. For practitioners, this endeavour includes but is not restricted to understanding the limits of their own personal professional competencies together with a willingness to identify and work toward rectifying relevant competency deficits. These are demanding requirements that imply a deep understanding of the turbulent and dynamic nature of practice, a recognition of the value of some form of critical self-reflection, and a resolve not to allow complacency to jeopardise future practice.

Sellman makes clear that an important consideration for any virtue-based account of professional conduct and activity is to recognise the situational constraints that can act upon and constrain the ability of professionals to act ethically. As Pitman (2012: 131) has argued, and as we have suggested above, the managerialism and marketisation of public professions such as teachers, healthcare professionals and social workers have created a 'hostile ground for growing phronesis' (see also Dixon-Woods et al., 2011). To neglect these factors is inherently problematic. Kinsella and Pitman (2012: 8) remind us that

> as the mechanisms of professionalization have been put in place, so too have the levels of prescription increased, thereby circumscribing the capacity of members to act autonomously in situations that demand the exercise of judgement. The 'danger' of calling for phronesis and holding practitioners accountable for practical wisdom in contexts that may not support it, and that actively mitigate against it, is that practitioners may

face a double bind, where they are blamed for a failure of agency at the personal level, when the issues may well be structural and systemic.

It is under such circumstances that moral and intellectual virtues – including the meta-virtue of *phronesis* – play a crucial role, enabling professionals to discern and deliberate about the correct course of actions given the *salient features at play* (Russell, 2009). Indeed, initial findings from a meta-analysis of professional virtues undertaken by the Jubilee Centre (Arthur et al., 2019: 5) indicate that the *'phronetic* professional is one that is posited to endorse both moral and intellectual virtues in conjunction with one another'. These initial findings suggest 'the importance of developing a *phronetic* character profile for the enhancement of perceived professional purpose. That is, one that encompasses a value for both moral and intellectual virtue simultaneously as opposed to in isolation of one another'. Importantly, moral virtues may be crucial for 'developing a sense of purpose that extends beyond the self to the community in which one works', but 'it is only when a moral compass is synergised with a valuation of the intellectual virtues, that professionals are likely to experience the greatest possible sense of professional purpose' (Arthur et al., 2019: 16). In other words, moral and intellectual values work together to guide right action and a deeper sense of professional worth.

Codes of Conduct and the Limitations of Rule

A core feature of professional occupations, then, is the ability to handle the ethical dilemmas and challenges faced within the workplace. Professional work is such that, given the complexity of their work and the challenges involved in delineating an ethically appropriate course of action, the professional cannot simply follow given guidelines or codes – particularly when ethical requirements conflict (e.g. when loyalty conflicts with honesty). So too, and given the complex nature and scope of professional activity, the professional must draw on a range of salient information – theories, practices, prevalent codes, relationships involved, potential outcomes – to discern the right course of action for the right reasons (see, for example, Fish and de Cossart, 2013). In certain circumstances, the complexity and challenges of their occupation may place professionals in situations where their actions may be both morally right and yet run counter to the requirements set out by the government and related agencies (Moore, 2015). As Carr (1999: 35) contends, 'responsible professional decisions must depend ultimately on the quality of *personal* deliberation and reflection'.

This is not to suggest, however, that the sort of practical wisdom needed for professional *phronesis* can be completely separated from the principles and rules that often characterise professional codes of conduct (Pellegrino and Thomasma, 1993). Having a clearly stated set of principles and rules brings a number of benefits in terms of educating new entrants to the profession, guiding professional conduct and providing those external to the profession (patients, clients, parents, pupils, etc.) with some understanding of what can be expected of the profession concerned (for an interesting discussion of codes of conduct in the military, see Mompeyssin, 2014). However, rules and codes of conduct can only help the professional so far and are insufficient for true ethical practice if they are not accompanied, interpreted and balanced by intellectual and moral character. In simple terms, where codes of conduct are too rigid, cultures of conformity can undermine professional autonomy and judgement; where codes of conduct are overly ambiguous, they offer professionals little by way of structure and guidance to act as a basis for their deliberations and choices.

American psychologist Barry Schwartz has spoken about the ways in which the dominance of external controls, such as rules and incentives, can actively *undermine* wisdom and judgement. According to Schwartz (2009),

> rules and incentives may make things better in the short run, but they create a downward spiral that makes them worse in the long run. Moral skill is chipped away by an over-reliance on rules that deprives us of the opportunity to improvise and learn from our improvisations. And moral will is undermined by an incessant appeal to incentives that destroy our desire to do the right thing. And without intending it, by appealing to rules and incentives, we are engaging in a war on wisdom.

Importantly for the focus of this book, Schwartz (2011) has also argued that the dominance of rules and incentives does not only limit professional wisdom but also serves to undermine professional motivation. He argues that

> the problem with relying on rules and incentives is that they demoralize professional activity, and they demoralize professional activity in two senses. First, they demoralize the people who are engaged in the activity. Judge Forer quits, and Ms. Dewey in completely disheartened. And second, they demoralize the activity itself. The very practice is demoralized, and the practitioners are demoralized. It creates people – when you manipulate incentives to get people to do the right thing – it creates people who are addicted to incentives. That is to say, it creates people who only do things for incentives.

The *phronetic* professional, then, is not guided solely by duty to codes external to their own intellect and morals or by externally driven incentives but rather conceives and applies their professional responsibilities by using their professional wisdom. This includes understanding codes of conduct but not conceiving them as the sole arbiter when dilemmas arise. As the author C. S.

Lewis (1985: 100; cited in Bohlin, 2005: 20) eloquently wrote in his *Letters to Children*:

> A prefect man would never act from a sense of duty; he'd always want the right thing more than the wrong one. Duty is only a substitute for love (of God and other people), like a crutch, which is the substitute for a leg. Most of us need the crutch at times; but of course its idiotic to use the crutch when our legs (our own loves, tastes, habits etc) can do the journey on their own.

Lewis' words remind us that sound professional conduct has an internal motivation and meaning – that is, it must come from the heart. It is for precisely this reason that many, if not all, professions are understood as vocations rather than simply occupations.

Focusing on the sorts of capacities frequently associated with professional *phronesis*, which include sensitivity, discernment, deliberation and reflection, signifies that the codification of professional conduct into a set of rules cannot be disentangled from the critical judgement of the professional. Indeed, the critical judgement of the professional is crucial if those rules are to be applied in practice and in a way that juggles the demands of the specific situation at hand (including where stated rules may be in conflict). Whether one subscribes to an Aristotelian notion of *phronesis* that separates ethical from technical practice or from a MacIntyrean approach that understands technical practice to have an ethical dimension, it remains that the ethical is core to professional practice (Cooke and Carr, 2014; Kotzee, Paton and Conroy, 2016; Kristjánsson, 2015b).

Conclusion

In this chapter we have surveyed the existing literature on the ethical dimensions of professions. As we have intimated in the chapter, it is not a question of *whether* professions such as medicine, law, nursing, social work, teaching and the military involve an ethical dimension but rather how this dimension is and should be conceived and enacted by these professions. While general approaches to professional ethics act as a significant starting point in responding to these latter questions, the nature, demands and realities of professional ethics are necessarily moderated by the particular profession at hand. In other words, while we might approach the general ethical dimensions of professions from a given framework (in the case of the Jubilee Centre, a broadly neo-Aristotelian one), it is also necessary to appreciate that the precise ethical demands that act upon doctors, nurses, lawyers, teachers and so on are likely to be framed and expressed in ways particular to those individual professions.

With this in mind, the focus of the next chapter is more specifically on ethics and the good soldier.

Notes

1 www.professions.com.au/about-us/what-is-a-professional
2 www.basw.co.uk/about-basw/code-ethics
3 www.lawsociety.org.uk/support-services/ethics/

2 Character, the Military and the Ethical Soldier

Introduction

Being a soldier, and being part of a professional military, necessarily involves ethical dimensions. In simple terms, soldiers are generally required to be courageous, be honourable and show integrity. So too, soldiers today are expected (and indeed may well expect themselves) to be respectful of others and to act with self-discipline in conduct. And the lack of these qualities is often highlighted when soldiers and armies fall short of their own standards and those expected of them by others. Furthermore, the character of soldiers is intimately connected with wider questions of military ethics, including just war theory and advances in technologies employed by contemporary militaries. Of course, how the character of soldiers should be understood, delineated and enacted remains the subject of much discussion – not least in terms of whether and how the desired qualities might be captured in official statements of Army Values and Standards and how these can be translated into ethical conduct given the complexity of being a soldier today.

On this basis, this chapter considers the extant literature on military ethics, with a specific focus on the army and on virtue-based understandings of the ethical soldier. The intention is not to provide an exhaustive account of the available literature but rather to provide an indicative sense of current thought in this area in order to contextualise and provide a foundation for the empirically focused chapters that follow. The chapter comprises four main sections. In the first, some initial statements are made regarding the military as the 'profession of arms'. In the second section, some key principles of general military ethics are introduced, including those of just war theory. The third section concentrates on virtue ethics and the military. The fourth section explores briefly existing literature on the changing nature of soldiering and military ethics.

The Profession of Arms

When one thinks of examples of a profession, perhaps those that most immediately come to mind include teaching, nursing, the clergy and the police,

DOI: 10.4324/9781003331209-3

among others. Yet the military has a long-standing, though not necessarily straightforward, status as a profession. Over time, the main purposes of a military have also been multifarious, including employing violence, protecting the state, acting domestically in times of emergency and intelligence gathering. Indeed, the proper role and standing of the military, including the relationship between the military, the state and society more generally, has remained a topic of some considerable academic interest. This interest is clearly identifiable within both the UK and the USA, the main focus of our present book (Dandeker and Freedman, 2002; Forster, 2006; Huntington, 1957; Janowitz, 1960; Morgan, 1994; Moskos and Wood, 1988; Moskos, Williams et al., 2000; Strachan, 2003).

In his classic book, General Sir John Hackett (1983: 9) referred to soldiers as members of the 'profession of arms', arguing that 'the function of the profession of arms is the ordered application of force in the resolution of a social or political problem'. The idea that a nation's army is a profession is not novel. In 1957, Samuel P. Huntington's *The Soldier and the State* defined officers, and all soldiers for that matter, as professionals. Huntington argued that 'the modern officer corps is a professional body, and the modern military officer is a professional' (1957: 7). Similarly, in his book *The Professional Soldier: A Social and Political Portrait*, Morris Janowitz (1960: 5–6) defined soldiers as 'professionals in violence'. Here, Janowitz argued that the professional offers a specialised service due to skills acquired by prolonged and intensive training, but more than this, members of a professional group 'develop a sense of group identity and a system of internal administration'. On this basis, and like law and medicine, the military is a profession. Others who have written about the military as a profession include Burk (2002), Challans (1999), Collins (2007), Miller (2004), Moten (2011), Robinson (2008), Snider et al. (1999), Snider and Watkins (2000, 2002), Snider (2000, 2010), Swain and Pierce (2019) and Wilson and Meese (2019).

Another important marker of a profession is its self-regulation by a code of ethics or conduct (Claypool et al., 1990; Collings-Hughes et al., 2022; Greenwood, 1957; Watkins, 1999; Wilensky, 1964). Higgs-Kleyn and Kapelianis (1999) note that the purpose of codes of conduct is the regulation of ethical behaviour in professionals. According to Gilman (2005), codes of conduct 'are the framework upon which professions are built', and in this sense codes of conduct can play a notable role in communicating the 'expected behaviour of those who practice within the profession' (Collings-Hughes et al., 2022). Additionally, codes are both informed by and speak to the societal good professions perform (Bayles, 1988; Oakley and Cocking 2002; Wolfendale, 2009). So far as actual militaries are concerned, it is usually the case (though not always) that the codes of conduct involved are titled values, as discussed in the introductory chapter to this book. We return to a more detailed discussion of military values and codes of conduct later in this chapter.

General Military Ethics

A key contention of this present book, the studies on which it is based, and of much of the literature engaged with in this chapter, is that soldiers are – and act as – moral agents who must navigate a range of tasks, activities and involvements (including at times taking lethal action). This view requires that soldiers be involved in understanding and regulating their own ethical conduct and what it means to be a good soldier within various ethical frameworks and practices. One of the defining theoretical principles impacting on military ethics over time has been just war theory and its distinction between *jus ad bellum* (concerned with the justice of going to war) and *jus in bello* (concerned with right action in war). Throughout history, most advanced civilisations, including the Aztecs, Babylonians, Chinese, Egyptians, mediaeval Europe and modern Western nations (Cox, 2018; Orend, 2002), have wrestled with what constitutes justifiable reasons for going to war and the morally acceptable conduct and actions during war. A central principle of just war theory is that, as Michael Walzer (2008), author of the hugely influential work *Just and Unjust Wars* (1977), notes, 'no government in high civilization . . . will send their young men into battle, to kill and be killed, without offering some justification for what they are doing'.

Long-standing principles of just war theory include, under *jus ad bellum*, the tradition established by the notions of just cause, right intention and public declaration by proper authority and, under *jus in bello* the notions of non-combatant immunity, right intention and non-use of prohibited weapons (Frowe, 2011, 2018; Orend, 2013). Regarding the former, *jus ad bellum*, the following six principles are central: (1) having a just cause in going to war; (2) having a legitimate or proper authority declaring war; (3) having the right intention; (4) having a reasonable probability of success; (5) proportionality, that is the morally weighted good achieved by the war outweighs the morally weighted bad that it will cause; and (6) entering the war is the last resort. There is now general agreement that resisting aggression by either self-defence or other defence is the only just cause for going to war, and this principle is supported through international law (Emerton and Handfield, 2018; Frowe, 2011; Orend, 2013). Article 2(4) of the United Nations (UN) Charter states that 'All Members shall refrain in their international relations from the threat or use of force against the territorial integrity or political independence of any state, or in any other manner inconsistent with the Purposes of the United Nations'. The UN Charter 51 takes this a step further, arguing that if a state is a victim of aggression, then the state has the right to protect itself:

Nothing in the present Charter shall impair the inherent right of individual or collective self-defence if an armed attack occurs against a Member of the United Nations, until the Security Council has taken measures necessary to maintain international peace and security. Measures taken by

Members in the exercise of this right of self-defence shall be immediately reported to the Security Council and shall not in any way affect the authority and responsibility of the Security Council under the present Charter to take at any time such action as it deems necessary in order to maintain or restore international peace and security.

Moving to *jus in bello*, the following six core principles are involved: (1) discrimination or non-combatant immunity; (2) proportionality, that is the foreseen but unintended harms must be proportionate to the military advantage achieved; (3) benevolent quarantine of prisoners of war; (4) non-use of prohibited weapons; (5) no 'mala in se' or 'methods evil in themselves'; and (6) no reprisals. Furthermore, the Law of Armed Conflict (LOAC), or International Humanitarian Law (IHL) as it is sometimes referred to, is a unique branch of international law which governs conduct in war, disarmament and war crimes. The LOAC seeks to strike a balance between the state's desire to defeat the enemy and the need to reduce suffering in war to the extent possible (Frederick and Johnson, 2015; Frowe, 2011; Peterson Legal, 2018; Solis, 2010).

The LOAC is derived from central planks of international law governing military ethics, such as the Geneva Conventions (concerned with the treatment of non-combatants, such as civilians and prisoners of war), the Hague Conventions (concerned with lawful weapons and targets) and the Chemical Weapons Convention (which forbids the use of all chemical weapons in war, even in self-defence). The LOAC has, itself, five main principles: military necessity, proportionality, unnecessary suffering, distinction and humanity. These formal laws provide both political and military leaders with guidance for their decision-making. Additionally, the LOAC provides guidance to members of the military in the midst of battle who 'may be faced in war with decisions of far greater moment than they would have encountered in civilian life' (Fisher, 2011: 84). Indeed, it is generally recognised that the LOAC has had a profound influence on the conduct of warfare (Frederick and Johnson, 2015; Solis, 2010). Additionally, most Western nations fighting in war follow Rules of Engagement (ROE), which dictate who and what can be targeted in specific war zones or 'theatres of operation', which frequently follow stricter rules than those in the broader sense of just war theory and the LOAC. ROE regulate the use of force by members of the military against deployed soldiers during peacetime and in wartime against an adversary (Boddens Hosang, 2020: 10; Duncan, 1999; Roach, 1983).

Virtue Ethics and the Military

In the previous section about military ethics, we discussed that both just war theory and the LOAC govern actions in war. By their very description, just war theory and the LOAC are largely rule-based constructs. Though in this sense these could be cast as broadly deontological in nature, it is the case that militaries, and those interested in military ethics, are also interested in the ideas of character and virtue. Ethos is the Greek word for character (Brahnam, 2009; Campbell, 1995; Halloran, 1982; Hannah and Avolio, 2011; Hannah and Jennings, 2013), while *Merriam-Webster* (2022) defines ethos as the distinguishing character, sentiment, moral nature or guiding beliefs of a person, group or institution. While militaries (and their subsequent branches like the army, navy and air forces) have mottos, they often refer to their mottos as their 'ethos' (Hannah et al., 2010). For instance, the US Army has 'The Warrior Ethos', which states, 'I will always place the mission first, I will never accept defeat, I will never quit, and I will never leave a fallen comrade' (HQ Army, 2011). As with other Western militaries, the US Army also refers specifically to the importance of character. A leading example of this is the mission statement of the United States Military Academy at West Point, which states:

> The United States Military Academy's mission is to educate, train and inspire the Corps of Cadets so that each graduate is a commissioned leader of character committed to the values of Duty, Honor, Country and prepared for a career of professional excellence and service to the nation as an officer in the United States Army.

While it has been said that military *culture* contributes most to individual motivation, ethical decision-making, ethos and character (Robinson, 2007), it is clear that virtue ethics informs the codes of conduct and moral education programmes of many armed forces (Robinson, 2008). As previously considered, many militaries have embraced a list of values, sometimes explicitly detailed as *virtues*, deemed to be representative of the expected conduct of members of the profession of arms (see Table 2.1 which replicates Table 0.1 for ease). The British Army lists six virtues within their *Values and Standards* (British MOD, 2022): Respect for Others, Courage, Loyalty, Integrity, Discipline and Selfless Commitment. The US Army (Department of Army, 2022) lists their particular virtues in the *Seven Army Values*. These virtues are represented in the acronym 'LDRSHIP' or 'leadership', which consists of Loyalty, Duty, Respect, Selfless Service, Honour, Integrity and Personal Courage. Many militaries listed in Table 2.1, such as the Australian Army (2022), British Army (British MOD, 2022), the Canadian Department of National Defense and Armed Forces (2022), the Norwegian Armed Forces (2022) and the US Army (Department of Army, 2022), have similar virtues in common, though each has their own distinctive interpretation of action as shaped by their respective histories and cultures. These similar virtues, such as Courage, Loyalty, Respect, Integrity and Service, might be considered 'martial virtues' or core virtues important to the profession of arms, with the intended capacity

Table 2.1 Values of Different National Armies

Australian Army's Core Values[1]	British Army's Values[2]	Canadian Forces Essential Values[3]	New Zealand Army's Values[4]	Norwegian Armed Forces'[5] Values	Spanish Land Army's[6] Values	US Army's[7] Values
Service	Courage	Duty	Courage	Respect	Valour	Loyalty
Courage	Discipline	Loyalty	Commitment	Responsibility	Spirit of Sacrifice	Duty
Respect	Respect for Others	Integrity	Comradeship	Courage	Discipline	Respect
Integrity	Integrity	Courage	Integrity		Companionship	Selfless Service
Excellence	Loyalty				Spirit of Service	Honor
	Selfless Commitment				Honour	Integrity
						Personal Courage

to inspire ethical conduct and good character – the hallmark of virtue-based understandings of the professions.

These 'martial virtues' noted, it also needs to be recognised that the virtues, character and ethical conduct of soldiers (and indeed of all military) operate in relation with the wider codes of conduct, theories of just war, international law and more specific rules of engagement considered above. As such, Robinson's (2008: 8) suggestion that focusing on soldiers' values and virtues can 'ignore the fact that the purpose of military ethics is not solely to produce soldiers who will be efficient but also to limit the use of force and to protect others from the power that soldiers wield' is interesting. In a similar vein, Wolfendale (2008) posits a possible tension in military ethics education between the cultivation of character and judgement on the one hand and educating awareness of codes of conduct and the need to follow rules of behaviour on the other.

This concern noted, character and virtues do clearly occupy an important place alongside (and at times intertwined within) more rule-based formulations of codes of conduct. The *Values and Standards of the British Army* are predicated on 'the type of person you [soldiers] are'.[8] The *Values and Standards* are established as defining 'who British soldiers are as individuals and what the British Army stands for as an organisation' and as 'the authoritative benchmarks against which we judge our conduct'. Indeed, meaningful accounts of the character and virtues of soldiers need to involve dispositional, contextual and legal factors and frameworks. This recognition is of importance for all professions but is of particular importance for the military given that military personnel – including soldiers – can be involved in some of the most difficult and humanly complex circumstances imaginable for character and ethical decision-making.

In certain situations, the choice between different actions may each result in ethically challenging outcomes (such as the loss of life, including on a large scale). In other words, there exist so far as the military is concerned ethically insoluble dilemmas (Robinson, 2015; Schulzke, 2013). In addition, the conduct of individual soldiers is also heavily impacted and delimited by the collective nature of armies (i.e. loyalty to the group) and also by the strictly regimented lines of authority and ranks. In other words, individual character, though vital, is not a lone force (Robinson, 2007; Sandin, 2007) and is impacted by culture, relationships with and to other soldiers, ethical climate and sound leadership. Indeed, and in line with the wider literature on virtue ethics, it should be noted that some commentators on military ethics have questioned the extent to which dispositions are the most important factor in guiding action. These commentators emphasise the significance of the situational and of the particular given *circumstances* of war over individual virtues and dispositions (Flanagan, 1991; Ross and Nisbett 1991; Tripodi, 2012). In response, those who advocate – as the authors of this current book do – that character and virtues *are* an essential factor in understanding and guiding ethical conduct take the view that situational and contextual factors are art of the relevant salient factors that operate alongside of, and impact upon, the character of those involved (in this case, soldiers). Here, and on a neo-Aristotelian-inspired virtue-based understanding of professional ethics as introduced in the last chapter, the meta-virtue of phronesis helps the moral agent (in this case, the soldier) to adjudicate the relevant factors involved and to discern the morally good course of action. Moreover, and of particular relevance to soldiers, some go so far as to suggest that 'virtue theory acknowledges that life often presents us with circumstances so challenging that few individuals possess the strength of character to overcome them, through their own resources' (Olson, 2014: 91).

Specific qualities pertinent to soldiers, including some not explicitly listed in Table 2.1 have also found expression, and have been discussed, in the literature. Courage is perhaps one of the more obvious military qualities (Kateb, 2004; Zavaliy and Aristidou 2014). According to Aristotle, military courage involves steadfastness in the face of death on the battlefield, and war offers an opportunity to express this. More recently, authors have differed as to precisely what might courage on the battlefield, and the demands of courage placed on soldiers, mean (see, for example, Olsthoorn, 2007). Important links between courage and comradeship have been made, often in the sense that these together engender a will to fight (Biggar, 2013; French, 2005; Shields, 1991; Shils and Janowitz 1975; Verweij, 2007). In terms of practical expression, the British Army Regimental System draws on the motivating forces of comradeship and small fighting units (Shils and Janowitz, 1975; Woodward, 2007). It must be noted that the bonds and loyalty forged in comradeship do require careful management, especially to the extent that such 'loyalty' might privilege obedience and/or is taken to excess (French, 2005; Kateb, 2004; Olson, 2014; Olsthoorn, 2011; Wolfendale, 2009). Once again, the meta-virtue of *phronesis* would seem to have a vital role in ensuring that excesses (and indeed deficiencies) are avoided and, crucially, can be brought to bear when virtues are in conflict (for instance, the virtues of honesty and loyalty).

The extant literature also contains some interest in honour and respect for others (Olsthoorn, 2005; Osiel, 2002; Robinson, 2007). According to French (2005: 5), honour is a crucial disposition for soldiers because 'warriors need a way to distinguish what they must do out of a sense of duty from what a serial killer does'. Honour has also been identified as helping soldiers take account of broader human concerns in their ethical thinking and conduct, in particular ideas of respect for human life and human dignity (Robinson, 2007). Respecting others is described by the British Army in their *A Soldier's Values and Standards*[9] as 'part of the trust that has to exist between you [soldiers] and your team mates', and 'means treating people decently', including 'civilians, detainees and captured enemy forces'. Resilience, to give another example, has been afforded some noted attention (cf. Boe, 2015; Jarrett, 2008; Meredith, Sherbourne et al., 2011). In addition, the US Army introduced a resilience programme, drawing heavily on work in positive psychology. The USA

Army Global Assessment Tool is a method used to assess factors that contribute to soldiers' resilience (Seligman, 2011).

The Changing Nature of Soldiering and Military Ethics

Historically, most nations' military purpose and strategy remain largely consistent – namely, to deliver prompt and sustained action, including violence when necessary, appropriate and proportionate through dominance in the air, sea and land. The phrase 'Mission First' underpins the military notion of placing the mission and the collective first over and above the individual. Over the last 100 years, however, the nature of the military and modern warfare has changed for a variety of social, technological and political reasons. Such changes have had implications for the personal qualities and characters of military personnel, as we examine in this section. For instance, during the First and Second World Wars, compulsory national service drew large numbers of men into the military ranks. In turn, women were called upon to undertake roles on the home front that supported the war, such as munitions and weapon production manufacturing and agricultural jobs. As such, 'whole societies were at war' (Dandeker, 1990: 101). Largely, though not necessarily exclusively, the military was focused on defence of attacks or invasion (Battistelli, 2000).

However, certain long-held assumptions about the role of the military and of military engagement are being rethought. Although some question the extent of change from previous times (Goulding, 2000; Shaw, 2005), many have raised the idea that we are in a new, even post-modern military, era (Battistelli, 2000; Eco, 1991; Forster, 2006; Hajjar, 2014; Kiszely, 2009; Lucas, 2010; Manigart, 2005; Micewski, 2005; Moskos et al., 1994; Moskos and Burke, 1994; Moskos, 2000; Snider, 2000, 2010; Williams, 2008). The use of the military to escalate force has given way to war deterrence and de-escalation through Peace Support Operations (PSO) and Military Operations other than War (MOOTW) such as the Responsibility to Protect (R2P), with most militaries now participating in peacekeeping and humanitarian aid (Caforio, 2006; Forster, 2006; Lazar, 2017; Orend, 2013). R2P is the principle that supports intervention into a state that has failed (regardless of intention) to reasonably protect a group of its citizens' physical security rights (Bellamy and Dunne, 2016; Bellamy and Luck, 2018; Burkhardt, 2017; Orend, 2013). In addition, many countries have deployed their military, including armies, as part of wider responses to the emergence of border/cross-border security concerns, including illegal migration/immigration, drug smuggling and terrorism (Edmunds, 2006).

Furthermore, many Western democratic countries have become all voluntary forces, including ending conscription and periods of national service (and even some countries which still have national service in principle do not enforce participation). In both the UK and the USA, the movement to all voluntary forces served to enhance professional identities coupled with all-smaller, highly skilled, flexible personnel (Edmunds, 2006). Notably and importantly, militaries more generally and armies more specifically have become increasingly diverse, particularly in terms of gender and sexuality. Additionally, civilians now fill many positions originally held by members of the military (Barnes, 2016; Caforio, 2006; Forster, 2006; Hajjar, 2014; Snider, 2000). Even in terms of the battlefield, there has been an increase of civilian employees of private military and defence contracting companies (Barnes, 2016). For soldiers, interactions are more varied as a result, including with military personnel from other friendly nations, aid workers, irregular forces, military-contractors, third country nationals and local civilians (Fletcher, 2004). Even in terms of the enemy combatants, large uniformed enemy units have given way to diffuse, decentralised, smaller units often without insignia, country affiliation and uniforms at all (Edmunds, 2006; Kaldor, 1999; Kiszely, 2009; Lucas, 2010). This latter shift is particularly important in relation to non-state forces (Fisher, 2011; Lazar, 2017; Orend, 2013).

A further, and increasingly significant factor driving change in military engagement, warfare, conduct and ethics warfare is the so-called Revolution in Military Affairs, including precision bombing and advanced communication and intelligence systems (see Kirkpatrick 2015a, 2015b and Sparrow 2015 for a recent discussion of this in relation to virtues). Technologies that were not available in the modern military era are ubiquitous in military action in the post-modern era (Lucas, 2010). The emergence of military technologies includes lethal autonomous weapons (LAWS) (Bieri and Dickow, 2014; Horowitz, 2019; Krishnan, 2009; Meier, 2016; Roff, 2014; Sayler, 2020; Stratman, 2018; Surber, 2018; Verbruggen, 2019), innovations in nanotechnology (Edwards et al., 2017; Molestina et al., 2020; Nasu and Faunce, 2009; Pitschmann and Hon, 2016; Ramsden, 2012; Tate et al., 2015), cyber warfare (Ducich, 2018; Dunlap, 2016; Glazier et al., 2017; Ramsey, 2018; Solis, 2014) and the use of nonlethal weapons (Coleman, 2015; Jauchem and Cook, 2007; Levine and Rutigliano, 2015; Lewer and Davison, 2005; Orbons, 2012; Orend, 2013: 134).

For many people, improvement in technology equates to 'progress', which leads to a dangerous assumption that technological advancement is necessarily good. Additionally, the speed of military technology outpaces and exceeds our understanding of the ethical, let alone political and legal, ramifications of using them in war (Chapa, 2022: 180; Kaag and Kreps, 2014, 11; LiVecche, 2021: xi). This means that gaining proficiency and expertise in military technology is a persistent and incessant objective. If the factual and conceptual understanding of advances in military technology cannot keep pace, ethical issues will undoubtedly result. Use of technology on the battlefield may increase proficiency and precision, but it also requires a marked increase in decision-making. The proliferation and increased deployment of high-speed and destructive military technology can result not just in increased physical damage to infrastructure and human bodies but also in moral wounds for those who did not make the morally correct decision in the use of the technology (LiVecche, 2021: ix) – in other words, in 'moral injury'. Moral injury occurs when an agent participates in a situation (such as witnessing, failing to prevent or perpetrating the act) that transgresses their deeply held moral beliefs and expectations about humanity (Chapa, 2022: 98; Frankfurt and Frazier, 2016; Griffen et al., 2019; Litz et al., 2009; LiVecche, 2021: 25; Sherman, 2015). In other words, human beings who are exposed to traumatic events that violate their moral values as an actor or observer may experience severe distress and functional impairments. In those moments of quick decision-making on the battlefield in the midst of military technology, military personnel may actually use the virtues that they have habituated in order to prevent or lessen immoral action or inaction, ultimately preventing moral injury.

Conclusion

This chapter has provided an overview of existing literature on military ethics, paying particular attention to the place of character and virtues in relation to the various themes included. The examination provided has sought to draw out how matters of ethics and character feature in relation to wider questions in military ethics, as well as drawing out the key qualities – or virtues – commonly identified as crucial for being an ethically good soldier. Notably, and perhaps in some way differently from other professions, the virtues of soldiers are routinely and explicitly cited in various statements of values and standards. Notable here is the bringing together of qualities one might more readily associate with combat – such as courage, loyalty, discipline and service – and those that speak the social and relational nature of military service, including soldiering – such as respect for others and integrity. On this basis, in the next two chapters the focus of the book moves now to present and analyse the empirical data gathered through the *Soldiers of Character* study to explore how officers in the British Army reason ethically as well as how they perceive and experience values, virtues and character.

Notes

1 www.army.gov.au/our-people/our-values-contract
2 www.army.mod.uk/who-we-are/our-people/a-soldiers-values-and-standards/
3 www.canada.ca/en/department-national-defence/services/benefits-military/defence-ethics/policies-publications/code-value-ethics.html
4 www.nzdf.mil.nz/army/
5 www.forsvaret.no/en/about-us/missions-and-values/values
6 https://ejercito.defensa.gob.es/en/personal/valores/index.html?__locale=en
7 www.army.mil/values/
8 www.army.mod.uk/who-we-are/our-people/a-soldiers-values-and-standards/
9 www.army.mod.uk/who-we-are/our-people/a-soldiers-values-and-standards/

3   Soldiers of Character

Ethical Dilemmas and Character Strengths

Introduction

This chapter presents and analyses empirical data drawn from the Jubilee Centre Project – *Soldiers of Character* (Arthur et al., 2018). More specifically, the chapter explores the ethical reasoning of soldiers at different stages of their careers in the British Army in relation to a set of four moral dilemmas. In addition, the chapter considers data that draws on soldiers' self-reports of virtues important to how they are obtained through the VIA-IS-EI measure. The chapter comprises four main sections. In the first, neo-Kohlbergian approaches to measuring ethical reasoning and the core idea of intermediate concepts are explained. In this first section, the Defining Issues Test and the ICM are introduced, including army-specific ICMs. In the second section, we introduce the four dilemmas that were part of the AICM used in the *Soldiers of Character* (Arthur et al., 2018: 15) study. The third section moves to an examination of the moral dilemma responses and scores. In the fourth and final section, we consider respondents' self-reported virtues, including how these relate to those found in sets of values of national armies and those prioritised by professionals in other recent studies conducted by the Jubilee Centre.

Neo-Kohlbergians, Moral Reasoning and Intermediate Concepts

As it relates to the study of moral reasoning, a neo-Kohlbergian approach builds on the more widely known Kohlbergian theory of staged moral development to provide a more intricate and gradual model of staged moral development (Mechler and Thoma, 2013). In contrast to Kohlberg's fixed stages, the neo-Kohlbergian approach uses schemas as a type of reasoning in order to explain how people 'understand, organize, and prioritize moral content such as societal norms, systems, and organizations' (Rest et al., 2000: 384–386; Thoma, 2014). A schema is a structure of general knowledge that resides in a person's long-term memory (Rest et al., 1999a) and is built and strengthened as people encounter comparable and repeated situations and life

DOI: 10.4324/9781003331209-4

scenarios. The schema allows the agent to apply the prior knowledge to new situations, helping the individual to understand, reason and judge the situation based on that prior knowledge (Parsons, 2021). Cantor (1990: 738) points out that schemas that are used more often become more readily accessible and, as such, reinforces their use in moral dilemmas in the future. Further, and according to Kristjánsson (2017), the schema approach corresponds with the neo-Aristotelian notion of 'habituated virtue'.

For neo-Kolbergians, there are typically three types of schema that lead to a moral judgement: bedrock schema, codes of conduct and intermediate concepts (Bebeau and Thoma, 1999; Thoma, 1986, 2014). Bedrock schemas are a broad and wide-ranging way of interpreting moral dilemmas. Conversely, codes of conduct are concrete, prescriptive and do not require interpretation. Moral judgements at the intermediate concepts schema are more abstract, and require more interpretation, than codes of conduct (Parsons, 2021). Intermediate concepts apply to a whole range of moral dilemmas that are not tied to a single event or cause and are wider than codes of conduct but more specific than the abstract moral schemes of Kohlberg. Moreover, it has been suggested that it is the intermediate concepts schema that accord more accurately with the variety of moral dilemmas professionals face in their respective fields and, by extension, should find greater expression in professional ethics education (Arthur et al., 2018; Arthur and Earl, 2020; Bebeau and Thoma, 1999; Thoma, 2006, 2014; Thoma et al., 2013).

Neo-Kohlbergian Measurements of Moral Reasoning

Neo-Kohlbergian scholars have created and developed several measurements to assess moral reasoning and moral development based on the intermediate concepts schema. Here we focus on two: the Defining Issues Test (DIT) (Rest et al., 1999a, 1999b; Thoma, 2006, 2014) and the ICM (Arthur et al., 2018; Bebeau and Thoma, 1999; Parsons, 2021; Thoma, 2014; Thoma et al., 2013; Turner, 2008).

The Defining Issues Test

Developed by James Rest and colleagues and consisting of six moral dilemmas, the DIT was first introduced in 1974. The DIT identifies that, when deliberating on macro-moral situations or dilemmas, there are three schema which become progressively more sophisticated in moral reasoning: preconventional, which focuses on personal interests; conventional, which focuses on maintaining norms; and post-conventional thinking (Narvaez, 2005; Rest et al., 1999a, 2000; Thoma, 2014). Developed during childhood, in the pre-conventional schema, the individual makes decisions in moral dilemmas based on what may be gained or lost personally. The conventional schema takes into account societal moral considerations and norms in discerning the action taken in response to a moral dilemma. In the post-conventional schema, moral duties and obligations are based on shared ideals, open to debate and based on the experience of the community (Parsons, 2021). In addition, the post-conventional schema has four necessary elements: the primacy of moral criteria, an appeal to an ideal, sharable ideals and full reciprocity.

When taking the DIT, the individual is asked to respond to each of the moral dilemmas by ranking a series of 12 options in order of importance to solving the moral dilemma. When completed, the individual receives a 'P-score' (a principled morality score based on their ranking of post-conventional options), resulting from their ranking across all six moral dilemmas. The DIT has been used to evaluate a variety of moral issues in a variety of disciplines (Bunch, 2005; King and Mayhew, 2002). Studies have evidenced that most populations have a significant number of individuals at different moral stages and schemas (Bunch, 2005; King and Mayhew, 2002; Parsons, 2021; Turner, 2008) and that respondents consider the morality of an action within a moral dilemma differently based on the schema in which respondents sit (Parsons, 2021; Rest et al., 1999a, 1999b; Turner, 2008). The DIT2 test has been developed in an effort to modernise the moral dilemmas and their options while also reducing the moral dilemmas from six to five to improve the measure's validity.

The Intermediate Concept Measure

The ICM bears some similarities with the DIT/DIT2. Both start with a moral dilemma, and both give a variety of choices for actions and a variety of reasons for the chosen actions in response to the moral dilemma. However, the ICM differs from the DIT in several important ways (Bebeau and Thoma, 1999; Parsons, 2021; Thoma et al., 2013; Thoma, 2014; Turner, 2008). First, the dilemmas in an ICM focus collectively on a specific population or profession, and dilemmas specific to those populations (soldiers, dentists, teachers, college students and adolescents, to name a few) rather than on abstract moral dilemmas as used in the DIT/DIT2. Second, an ICM offers both multiple options for actions and multiple options for reasons for the chosen action. Third, the actions and reasons chosen are 'scored' against a 'key' of answers created by an expert panel comprised of experienced members of the focused population or profession.

The main use of ICMs has been to examine the moral reasoning of professionals, including measuring the effectiveness of professional ethics training. The first ICM was developed by Bebeau and Thoma (1994, 1998a, 1998b, 1999) as a measure of ethics instruction for the dental profession. The Dental Ethical Reasoning and Judgement Test (DERJT) measured dentist professionals' ability to reason through moral dilemmas using moral concepts that applied specifically to the field dentistry. Other ICMs have been used to examine the moral reasoning of adolescent groups (see, for instance, Arthur et al., 2015a, 2015b; Parsons, 2021; Thoma et al., 2013; Thoma, 2014; Turner, 2008). Specifically, the Intermediate Concepts Measure for Adolescents (AD-icm) was designed as a way to assess the effectiveness of character education programmes in improving moral reasoning in adolescents. In addition to the AD-icm, the Ad-ICM(UK) was used to assess the moral judgement of young people in UK schools (Arthur et al., 2015; Walker et al., 2017). In addition to adolescents and the dental profession, ICMs have been developed to assess the effectiveness of professional ethics instruction in other occupations such as medicine (Arthur et al., 2015; Pinijphon, 2009), teachers (Kerr, 2021), pharmacists (Roche et al., 2014) and – as is our focus in this book – of military officers/cadets (Arthur et al., 2018; Arthur et al., 2020; Parsons, 2021; Turner, 2008; Walker, 2020).

ICMs Specific to the Army

As explained in the introductory chapter, the *Soldiers of Character* project developed and employed the AICM to examine the character strengths and ethical reasoning of soldiers at different stages of their careers in the British Army. Before providing details of the dilemmas used in the AICM and presenting key findings, it is worth briefly noting two other ICMs designed specifically to measure the effectiveness of ethics training on the moral reasoning of army officers and military academy cadets. The first, the Army Leader Ethical Reasoning Test (ALERT), was designed by US Army Lieutenant Colonel Michael Turner (2008) and was aimed specifically at cadets at the United States Military Academy at West Point (West Point). ALERT consisted of seven moral dilemmas set in a US Army-specific context, each followed by two sections. The first section involved the available action choices in regard to the dilemma and the second a number of available reasons, or justifications, for the action choices selected by the respondent using a five-point Likert-type scale. Importantly, the actions and justifications were scored as either acceptable/unacceptable or important/not important based on a key developed by an expert panel of senior military officers, thereby enabling the scoring and comparing of respondents' choices. Comparing the respondents' choices with the scoring key generated three main summary scores: (1) the percentage of times a respondent's action choice rankings (for both good and bad items) matched the experts' rankings; (2) the percentage of times a respondent's reasoning for their action choice (for both good and bad items) matched the experts' rankings; and (3) a total score that combines the choice and reasoning for the action.

A second ICM designed, the Army Reasoning and Ethical Training and Education Test (ARETE), also seeks to measure the effectiveness of ethics training on the moral reasoning of Army officers and military academy cadets.

The ARETE was adapted from Turner's (2008) ALERT by one of the authors of this book, US Army Major Scott Parsons (2021), who designed the ARETE to measure the moral reasoning and judgement of West Point cadets through their choices for actions and reasons for actions in a series of army-specific moral dilemmas. The ARETE consisted of six moral dilemmas set in a US Army-specific context, each followed by two sections. Like Turner's ALERT, ARETE's first section involved the available action choices in regard to the dilemma and the second a number of available reasons, or justifications, for the action choices selected by the respondent using a five-point Likert-type scale. Again, the actions and justifications were scored as either acceptable/unacceptable or important/not important based on a key developed by an expert panel of senior military officers, thereby enabling the scoring and comparing of respondents' choices. Comparing the respondents' choices with the scoring key generated three main summary scores: (1) the percentage of times a respondent's action choice rankings (for both good and bad items) matched the experts' rankings; (2) the percentage of times a respondent's reasoning for their action choice (for both good and bad items) matched the experts' rankings; and (3) a total score that combines the choice and reasoning for the action.

The AICM Dilemmas

All of the studies focusing on professions conducted by the Jubilee Centre have involved respondents engaging with a number of ethical dilemmas. In the *Soldiers of Character* project, four ethical dilemmas were developed. As explained in previous chapters of this book, and as outlined in the *Soldiers of Character* (Arthur et al., 2018: 15) report, the purpose of engaging respondents with the dilemmas was to explore 'intermediate concepts', which 'are assumed to lie between so called "bedrock" schemas of moral reasoning (self-interest; maintaining norms; and post-conventional schemas) and specific contextual norms (such as professional codes)'. The four dilemmas were adapted from the ALERT. The original research team modified the dilemmas to ensure they reflected the values and context of the British Army (the process for this modification was set out in the introduction to this book). In this section, a brief overview of the four moral dilemmas used in the AICM is provided. To see the full moral dilemmas and the action and reason choices, see Appendix 1.

Dilemma 1: When undertaking a non-time sensitive resupply tasking, *Captain Metcalf* needs to make a decision about how to respond to an injured local Somalian, who is a British Army informant and who is surrounded by an unpredictable crowd.

Dilemma 2: During major combat operations, *Major Smith* has to decide how to treat two captured enemy soldiers who may know the location of two of

Major Smith's soldiers who themselves have been captured by the enemy and are under immediate threat.

Dilemma 3: *Lieutenant Colonel Milgram* has to decide how to respond to an investigation into his unit, who were using nonlethal force that might have accidently killed two Iraqi civilians. Milgram knew about the use of several nonlethal tactics and had condoned them. However, Milgram never authorised the tactic that was used on the two possible dead civilians, which involved throwing them into a river.

Dilemma 4: *Lieutenant Jacobs* has to decide how to respond to a fellow male officer and friend who is fraternising with a female soldier, contrary to army rules.

After reading each moral dilemma, the British Army officers and cadets were asked to rate a series of possible responses to the situation described in each dilemma on a five-point Likert-type scale:

1 – I strongly believe this is a GOOD choice.
2 – I believe that this is a GOOD choice.
3 – I am not sure.
4 – I believe this is a BAD choice.
5 – I strongly believe this is a BAD choice.

After rating the possible action choices, the officers and cadets were asked to select the two best actions of the ones they had rated (best solution to the dilemma and the second-best solution to the dilemma). The officers and cadets were then asked to select the two worst action choices of the ones they had rated (worst solution to the dilemma and the second-worst solution to the dilemma). Next, the officers and cadets were asked to rate possible reasons, and the importance of the reasons, behind the decision on what ought to be done in each dilemma, again on a five-point Likert-type scale:

1 – I strongly believe this is important.
2 – I believe this is important.
3 – I am not sure.
4 – I believe this is not important.
5 – I strongly believe this is not important.

After rating the importance of possible reason choices (justifications) when making up their minds to act to the situation described in the moral dilemma, respondents were asked to select the two most important reasons (justifications) of the ones they had rated (best reason to act in the dilemma and then the second-best reason to act to the dilemma). Then the cadets were asked to select the two least important reasons (justifications) of the ones they had rated (worst reason to act in the dilemma and then the second worst reason to act in the dilemma).

Moral Dilemma Responses and Scores

Overall Responses

In Table 3.1 the mean percentages for the primary ICM indices are set out. When these findings are considered, it is evident that officers scored well over 50 per cent (M = .65).[1] This indicates that as a collective the responses of the officers were quite close to the expert panel judgements regarding the four dilemmas, in terms of both what actions should be taken and the reasons selected for those actions. Results were evenly distributed across percentiles (25th = .57; 50th = .68; 75th = .76).

The data displayed in Table 3.1 also suggests that the officers were closer to the expert panel judgements when selecting the best action (M = .66) and worst action (M = .73) than when selecting the best justification (M = .62) and worst justification (M = .60). Given that the mean percentage scores for selecting the best and worst actions and for selecting the best and worst justifications were well above 50 per cent, it would seem that officers are not weak in identifying justifications, but rather that they found selecting actions easier than justifications (and selecting good justifications a little easier than bad justifications). Within-subject differences on the four subscales of the AICM were tested by the project team through a repeated measures ANOVA (Analysis of Variance). The results indicated a significant main subscale effect using the Greenhouse-Geisser correction for the absence of sphericity (F (2.63, 620.17) = 24.44; $p < .001$; $\eta_p 2 = .094$).[2] All subsequent repeated measure ANOVAs were subject to the same procedures to test and correct for the absence of sphericity. Inspection of the individual contrast between means confirmed that action choices had higher means than justification choices (Arthur et al., 2018).

Considering the data in Table 3.1 further also gives additional insights into the AICM findings in relation to gender. Analysis of this data found that female officers (M = .69) had higher overall Total ICM means than male officers (M = .64) (F(1,235) = 4.85, $p < .05$, $\eta_p^2 = 0.020$). As highlighted by the

Table 3.1 AICM (%) Scores for Officers by Gender

Variable	Categories	Sample size	Total ICM	Subscales			
				Action choices		Justification choices	
				Best	Worst	Best	Worst
		238	0.65 (0.14)	0.66 (0.22)	0.73 (0.17)	0.62 (0.23)	0.60 (0.17)
Gender	Male	187	0.64 (0.14)	0.64 (0.22)	0.71 (0.18)	0.62 (0.23)	0.60 (0.17)
	Female	51	0.69 (0.11)	0.71 (0.21)	0.78 (0.17)	0.64 (0.23)	0.63 (0.17)

Note: Standard deviations are shown in parenthesis.

original project team, it is interesting to note that the gender differences set out in Table 3.1 are smaller than differences between gender found by previous studies employing moral dilemmas or ICMs, in which females outperform males (Thoma, 1986; Thoma et al., 2013; Walker, 2006).

Analysis of the four AICM subscales found a moderate between-subject main effect for gender (F(1,235) = 4.85, p < .005, η_p^2 = .020). Where choosing best and worst action choices was concerned, female AICM scores (M = .71 and M = .78) were 7 percentage points higher than male scores (M = .64 and M = .71). Turning to selecting best and worst justifications, the difference between females (M = .64 and M = .63) and males (M = .62 and M = .60) was insignificant, indicating that while female officers were somewhat better than their male counterparts in selecting action choices, female and male officers were equally adept at choosing justifications.

Responses for the Individual Dilemmas

We turn now to the findings with regard to each of the individual dilemmas. Chart 3.1 shows the overall mean scores for each of the four dilemmas – Metcalf, Smith, Milgram and Jacobs. Of the four, the highest mean score was for the Smith dilemma (M = .74), which focused on torture/aggressive methods. The second highest was for the Milgram dilemma (M = .70), which involved soldiers' use of non-authorised methods. The third highest was for the Jacobs dilemma (M = .61), which was concerned with fraternisation, while the lowest mean score was for the Metcalf dilemma (M = 53), which focused on the injured local Somalian. These findings were indicated by a

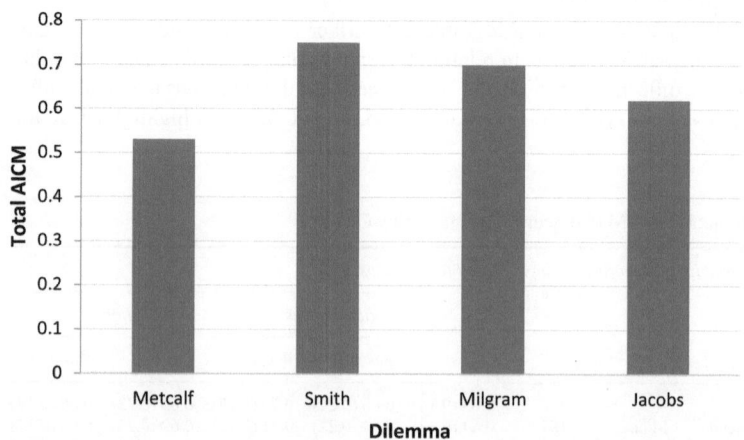

Chart 3.1 AICM (%) Scores by Dilemma

significant repeated measures ANOVA with dilemma as the within-subjects factor $F(3,705) = 30.030; p < 001, \eta_p^2 = .11$.

This main effect was conditioned by a gender by dilemma interaction effect $F(3,705) = 2.857; p < 05, \eta_p^2 = .012$. Scores per dilemma and gender are shown in Chart 3.2.

When responses to the individual dilemmas were considered by gender, analysis found that female respondents scored more highly than male respondents for three of the four dilemmas – namely, Metcalf, Smith and Jacobs. The highest scores for both male and female officers were for the Smith dilemma, while the lowest for both were for the Metcalf dilemma. This said, and again as Chart 3.2 demonstrates, aside from the Jacobs dilemma (dilemma 4), it is the similarities between gender rather than the differences that stand out.

Charts 3.3 and 3.4 set out the mean AICM scores by (i) held rank and by (ii) held rank and branch of service respectively. When the data were considered in relation to the army rank held, majors (M = .67) and cadets (M = .68) scored higher on the dilemmas than lieutenants (M = .65) and captains (M = .63). These descriptive differences were not statistically significant. However, when the officers were divided into infantry/artillery on the one hand and all other branches of service (i.e. non-infantry/artillery) on the other, a comparison of the total AICM scores for each of the two groups of officers evidenced a significant interaction effect $(F(2,205) = 3.088 \ p < .05, \eta_p^2 = .036)$. As set out in Chart 3.4, infantry/artillery officers scored higher than non-infantry/artillery officers at the lieutenant and captain/major[3] levels but not at the officer cadet level. Other than these differences, total AICM scores followed similar patterns by rank for infantry/artillery officers versus other branches of service.

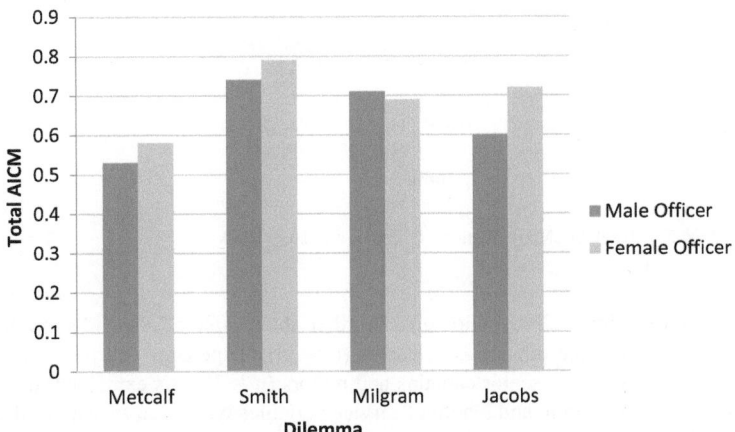

Chart 3.2 AICM (%) Scores for each Dilemma by Gender

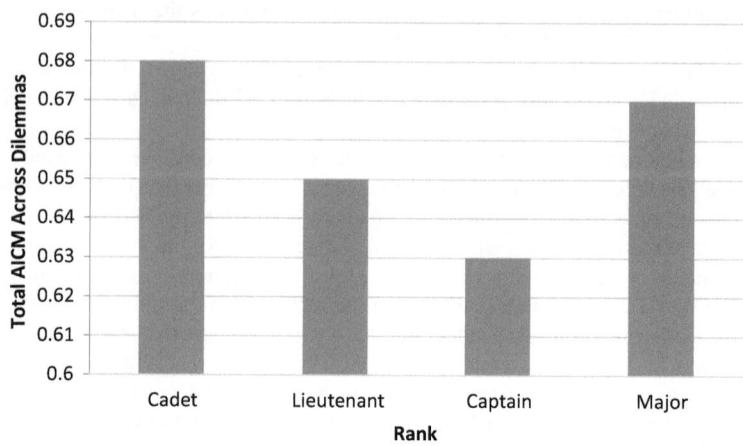

Chart 3.3 Held Rank and AICM Scores across Dilemmas

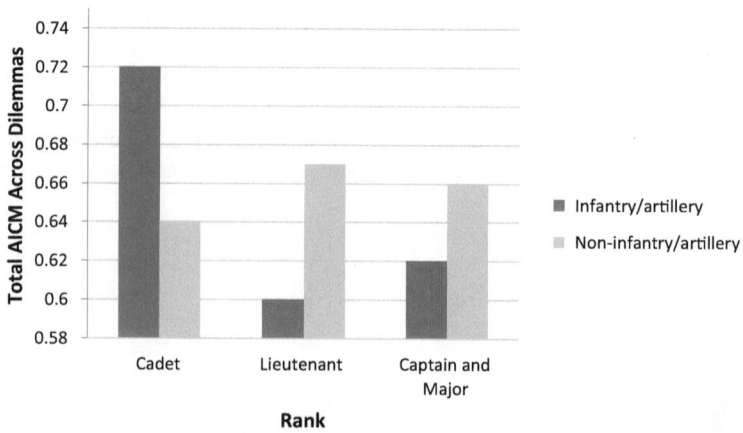

Chart 3.4 Total AICM by Branch of Service and Held Rank

To examine the interaction effect further, the sample of respondents were divided into three rank groups: cadets (n = 76); lieutenants and junior captains (n = 93); and senior captains and majors (n = 73). As explained in the original report, rank and length of service variables were used to achieve this by dividing the captain rank group into senior (6 or more years' service) and junior holders of this rank (1 to 5 years' service). The significant interaction effect ($F(2,205) = 4.022\ p < .05$, $\eta_p^2 = .038$) persisted for this revised grouping

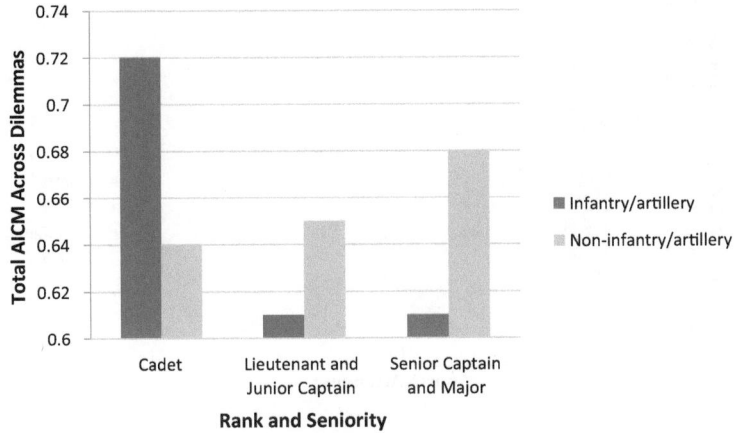

Chart 3.5 Performance by Rank and Seniority and Branch of Service

showing a dip in total AICM scores for infantry/artillery officers, as illustrated in Chart 3.5. In comparison, non-infantry/artillery officers as cadets scored well below their infantry/artillery counterparts, but the scores improved with seniority.

As the data were collected at specific career courses, officers were also grouped in this way to examine whether doing so generated different scoring patterns. The groups were Royal Military Academy Sandhurst – Cadets (RMAS) preparing to become officers (n = 49);[4] The Junior Officers' Tactics Awareness Course (JOTAC) – a course that prepares young British Army officers for their first role as a Captain (n = 81); and Captain's Welfare Course (CWC) – a course designed to make soldiers effective leaders (n = 81). Like previous rank-based groups, there was a significant interaction effect between the 'course' and infantry/artillery and non-infantry/artillery distinctions ($F_{(2,205)} = 4.559$ $p < .05$, $\eta_p^2 = .043$). However, this created a *lower* average result for JOTAC officers from branches of service other than infantry and artillery, scoring more closely to cadets from the same cap badge grouping (M = .64) than did lieutenants and junior captains. The total AICM by course and branch of service are illustrated in Chart 3.6.

Differences by Dilemma and Branch of Service

Each of the four dilemmas was analysed individually in terms of respondents' rank and seniority, again by branch of service (i.e. infantry/artillery and non-infantry/artillery). Of the four dilemmas, the first involving Metcalf and how to respond to an injured Somali surrounded by a large and unpredictable

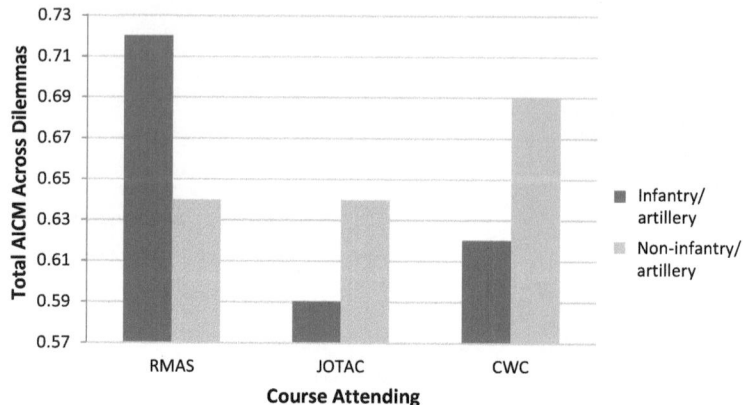

Chart 3.6 Total AICM by Course and Branch of Service

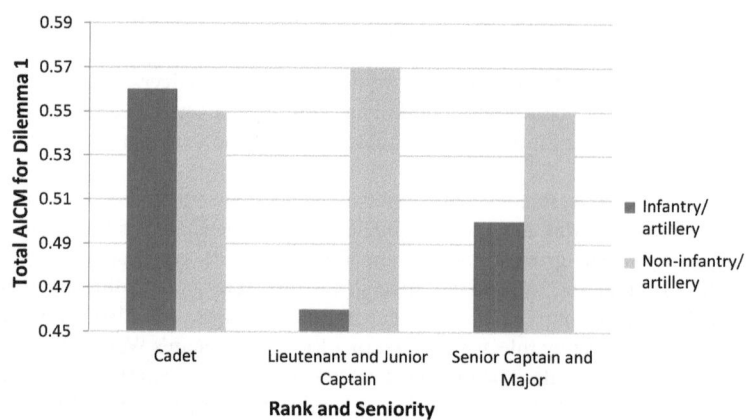

Chart 3.7 Dilemma 1 by Rank and Seniority and Branch of Service

crowd during a resupply task returned the lowest score overall (M = .54). The further analysis of this data, as captured in Chart 3.7, found that non-infantry/ artillery officers scored higher than infantry/artillery officers at the 'lieutenant and junior captain' and 'senior captain and major' levels but not at the cadet level. This finding was in keeping with broader scoring patterns. Chart 3.7 also highlights some evenness across the scores for dilemma 1 by non-infantry/ artillery officers, with the highest scores for lieutenants and junior captains (cadet (M = .55), lt. & jnr. capt.[5] (M = .57), snr. capt. & maj. (M = .54)). These scores for non-infantry/artillery officers stand in contrast to those for

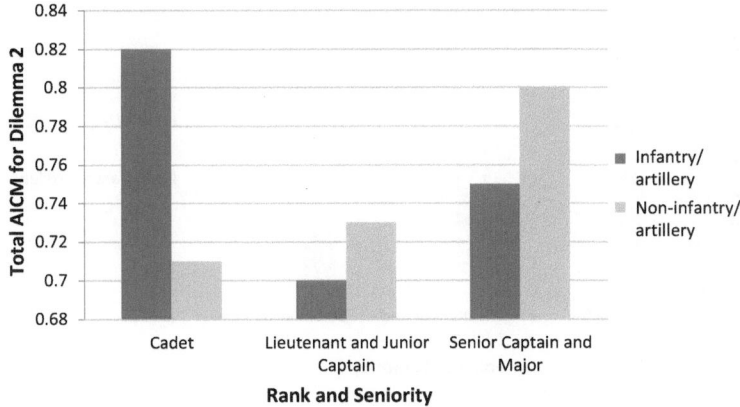

Chart 3.8 Dilemma 2 by Rank and Seniority and Branch of Service

infantry/artillery officers, where evenness across rank and seniority was not found (cadet (M = .56), lt. & jnr. capt. (M = .46), snr. capt. & maj. (M = .50)).

The second dilemma involving Smith concerned participants responding to time critical pressure from a sergeant major to get information about missing soldiers. This dilemma resulted in the highest overall scores (M = .75). As with dilemma 1, non-infantry/artillery officers scored higher than infantry/artillery officers except at the cadet level. At the cadet level, infantry/artillery scores were much higher than non-infantry/artillery (M = .82 versus M = .71). Once more, the scores for infantry/artillery officers were again lower at middle ranks (M = .70), with senior ranks at (M = .75). In contrast, the scores for non-infantry/artillery officers increased with each rank level (lt. & jnr. capt. (M = .73), snr. capt. & maj. (M = .80)). Chart 3.8 sets out the results for this dilemma.

Dilemma 3 (Milgram) concerned officers who were asked to respond to a possible criminal investigation about the use of non-authorised tactics by soldiers. The overall mean scores for this dilemma were high (M = .70). As Chart 3.9 demonstrates, for this dilemma – and in contrast to dilemmas 1 and 2 – infantry/artillery officers scored higher than non-infantry/artillery officers across all ranks. However, the scores for infantry/artillery officers across all ranks reduced slightly with seniority (cadet (M = .77), lt. & jnr. capt. (M = .72), snr. capt. & maj (M = .69)). Whilst the scores for non-infantry/artillery officers were lower than for infantry/artillery officers, these were reasonably consistent across the three rank groups (cadet (M = .69), lt. & jnr. capt. (M = .69), snr. capt. & maj. (M = .67)).

For dilemma 4 involving Jacobs, respondents considered a situation concerning how to react to a friend and fellow male officer who is having a

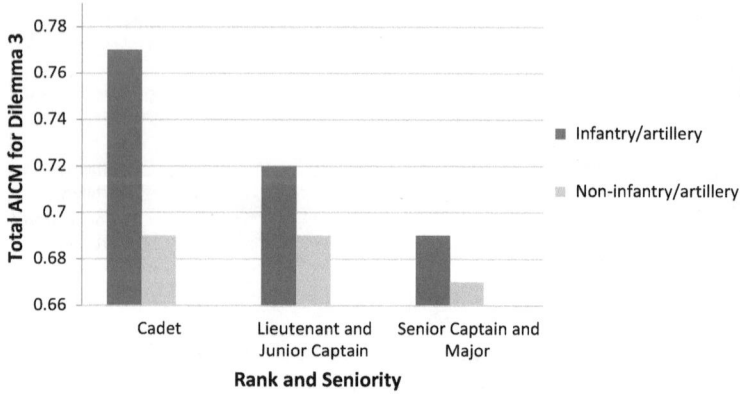

Chart 3.9 Dilemma 3 by Rank and Seniority and Branch of Service

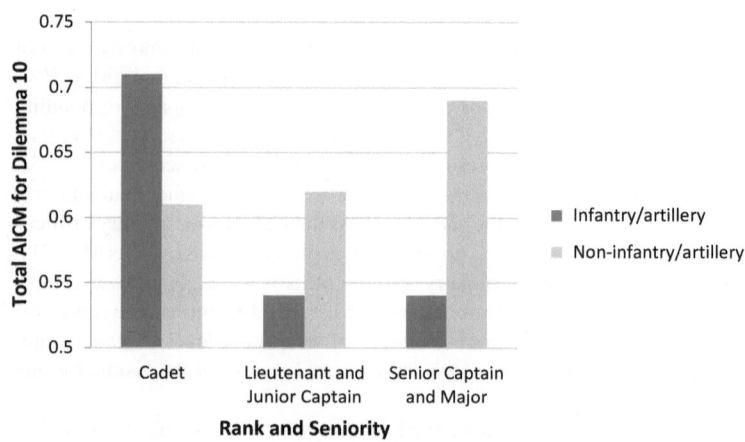

Chart 3.10 Dilemma 4 by Rank and Seniority and Branch of Service

relationship with a female soldier. The overall results for this dilemma were the third lowest ($M = .62$) of the four dilemmas. As Chart 3.10 sets out, for dilemma 4 infantry/artillery cadets had higher mean scores than non-infantry/artillery cadets ($M = .71$ versus $M = .61$), whereas for the scores for infantry/artillery lieutenants and junior captains and senior captains and majors (both $M = .54$) were lower than for non-infantry/artillery lieutenants and junior captains ($M = .62$) and senior captains and majors ($M = .69$).

Self-Reporting Measure – VIA-IS-E1

As explained in the introductory chapter, the *Soldiers of Character* project was also interested in officers' own character strengths, utilising the VIA-IS-E1 to obtain self-reports from respondents. The survey involved officers indicating how strongly they agreed or disagreed that each of 24 character strengths was an essential part of who they are in the world on the following scale: 'strongly agree' (5), 'definitely agree' (4), 'slightly agree' (3), 'neutral' (2), 'disagree' (1). Table 3.2 presents the ranked mean scores and standard deviations for the officers for all respondents and by gender. Average results for all officers are shown in Table 3.2 and are split by gender. Dominant strengths are at the top of the table and least dominant strengths are at the bottom. Ranked mean (average) scores and standard deviations for the officers are included.

Table 3.2 Values in Action Character Strengths

All (n = 226)	Male (n = 176)	Female (n = 50)
Teamwork (4.12, 0.78)	Teamwork (4.14, 0.79)	Perseverance (4.12, 0.90)
Honesty (4.12, 0.83)	Curiosity (4.13, 0.79)	Honesty (4.12, 0.80)
Curiosity (4.08, 0.80)	Honesty (4.12, 0.84)	Kindness (4.10, 0.93)
Fairness (4.07, 0.94)	Fairness (4.06, 0.94)	Fairness (4.10, 0.95)
Perseverance (4.04, 0.92)	Perseverance (4.02, 0.92)	Teamwork (4.08, 0.78)
Leadership (3.93, 0.82)	Leadership (3.98, 0.83)	Curiosity (3.94, 0.84)
Bravery (3.88, 0.88)	Bravery (3.90, 0.87)	Zest (3.88, 0.96)
Judgement (3.87, 1.02)	Judgement (3.89, 1.03)	Humility (3.86, 1.01)
Humour (3.84, 1.01)	Humour (3.85, 1.04)	Social Intelligence (3.84, 0.96)
Kindness (3.83, 0.89)	Perspective (3.82, 0.91)	Love of learning (3.82, 1.02)
Perspective (3.81, 0.94)	Kindness (3.75, 0.86)	Bravery (3.80, 0.93)
Humility (3.84, 1.01)	Humility (3.71, 1.02)	Humour (3.78, 0.91)
Social Intelligence (3.63, 1.12)	Love of learning (3.57, 1.09)	Leadership (3.78, 0.76)
Love of learning (3.63, 1.08)	Social Intelligence (3.57, 1.09)	Judgement (3.78, 1.00)
Zest (3.55, 1.04)	Creativity (3.46, 0.97)	Perspective (3.76, 1.04)
Gratitude (3.41, 1.04)	Zest (3.46, 1.05)	Love (3.72, 1.03)
Love (3.40, 1.13)	Gratitude (3.39, 1.01)	Gratitude (3.48, 1.15)
Creativity (3.39, 1.02)	Love (3.30, 1.15)	Hope (3.34, 1.27)
Hope (3.07, 1.16)	Forgiveness (3.02, 1.20)	Creativity (3.16, 1.15)
Forgiveness (3.01, 1.21)	Hope (2.99, 1.12)	Appreciation of Beauty (3.14, 1.26)
Appreciation of Beauty (2.98, 1.26)	Appreciation of Beauty (2.94, 1.27)	Forgiveness (2.96, 1.52)
Self-regulation (2.87, 1.11)	Self-regulation (2.87, 1.11)	Spirituality (2.96, 1.52)
Prudence (2.87, 1.20)	Prudence (2.86, 1.22)	Prudence (2.90, 1.13)
Spirituality (2.82, 1.48)	Spirituality (2.78, 1.47)	Self-regulation (2.86, 1.11)

The self-reporting character strengths data found that the overall top six reported strengths for all of the officers were teamwork, honesty, curiosity, fairness, perseverance and leadership. For male respondents, the top six were exactly the same, though in a slightly different order: teamwork, curiosity, honesty, fairness, perseverance and leadership. The top six for the female officers included five of the same six character strengths – perseverance, honesty, fairness, teamwork and curiosity – but included kindness rather than leadership.

These findings are interesting in that they are consistent with findings from other professions. Arthur et al. (2019) found that across a range of professions, seven virtues were ranked highest in both personal and professional domains. That similar character strengths to those highly rated across professions were emphasised by soldiers in the study lends further support to the idea that the military is a profession (Burk, 2002; Challans, 1999; Collins, 2007; Hackett, 1983; Huntington, 1963; Miller, 2004; Janowitz, 1960; Moten, 2011; Robinson, 2008; Snider et al., 1999; Snider, 2000, 2010; Snider and Watkins, 2000, 2002; Swain and Pierce, 2019; Wilson and Meese, 2019). Finally, and recognising that different values/virtues involved in these and the VIA-IS-E1 are not identical, it is interesting that the top six overall virtues that the cadets and officers chose in the Soldiers study were not in the MOD's Values and Standards (British MOD, 2022) nor in the US Army's (Department of Army, 2022) Seven Army Values.

Turning to the least dominant strengths, for the whole sample of officers these were spirituality, prudence, self-regulation, appreciation of beauty, forgiveness and hope. Once again, the male respondents' lowest six included the same six character strengths, this time in the same order. For the female respondents, the lowest six evidenced one difference. While self-regulation, prudence, spirituality, forgiveness and appreciation of beauty featured, creativity rather than hope was in the bottom six (though hope was the seventh lowest).

Further insights into the dilemma responses were gained by examining the self-reported character strengths in relation to the dilemmas. This analysis found that scores for the dilemmas were significantly and positively correlated with the self-reports of character strengths provided by the officers for the following character strengths:

judgement, $r = .168$ [.043, .295];
honesty, $r = .238$ [.103, .377];
bravery, $r = .152$ [−.013, .294];
perseverance, $r = .152$ [.027, .285];
fairness, $r = .253$ [.119, .393];
leadership, $r = .191$ [.041, .326];
prudence, $r = .170$ [.032, .295];
self-regulation, $r = .210$ [.088, .329];

In addition, scores were negatively correlated for one character strength – creativity, r = −.149 [−.269, −.023]. (All *p*'s < .005.) The relationships between AICM results and the strengths listed above were mainly positive. In other words, higher self-reported averages for these strengths corresponded to higher total AICM scores. This positive correlation, however, was not the case for *creativity*. Here, higher averages for this strength correlated with lower total AICM results.

Conclusion

In this chapter we have examined and explained, drawing on key data from the *Soldiers of Character* project, soldiers' responses to the series of ethical dilemmas contained within the AICM. As the chapter has suggested, the data showed a number of important findings, including that, when taken as a collective, officers' responses to the dilemmas were close to the judgements of the expert panel in regard to the actions selected and the justifications selected. Furthermore, deeper analysis revealed that officers were closer to the judgements of the expert panel when selecting actions than when selecting justifications. In addition, responses to the VIA-IS-EI suggested that certain virtues – namely, teamwork, honesty, curiosity, fairness, perseverance and leadership – were viewed by soldiers as most important to who they are in the world. Notably and importantly, the higher self-reported averages for strengths such as judgement, honesty, bravery and leadership were shown to correspond to higher total AICM scores. The focus now moves to Chapter 4, in which we consider data gathered through interviews with officers that explored their perceptions and experiences of values, virtues and character in the Army.

Notes

1 M=.65 represents 65 per cent agreement with expert panel judgements about acceptable and unacceptable options, including both action and justification selections – see the methods section for a full explanation.
2 It is standard practice to interpret partial eta squared (η_p^2) in the following way: 0.01 = small; 0.06 = medium; and 0.14 = large.
3 Captains and majors were combined in one group to achieve adequate sample size.
4 The lower sample size is because of cadets who did not yet know their future branch of service.
5 Abbreviations used here refer to lieutenant and junior captain (lt. & jnr. Capt.); senior captain and major (snr. capt. & maj.).

4   Soldiers' Perceptions of Character

Introduction

As explained in the introductory chapter, the *Soldiers of Character* study involved semi-structured interviews with 40 officers that explored their perceptions and experiences of values, virtues and character in the Army. The interviews focused on a range of matters. These included (1) soldiers' motivations for joining the Army; (2) their views on the extent of transfer of Army Values and Standards across professional and personal lives; (3) the personal qualities and character strengths an ideal officer of their own rank might have; (4) the personal qualities or strengths most important to them in their current role as an Army officer; and (5) the pressures or barriers that make it difficult for them – or others like them – to act ethically. In the original report, the data from the semi-structured interviews were examined in relation to two groups based on the AICM measure: (a) the 10 lowest scoring officers and (b) the 10 highest scoring officers. Here, however, we look across all of the interviews to examine the data for each of the five matters above in turn. The data suggests that across and within each of the five themes examined, officers held various perceptions of virtues and character in the Army and had a range of experiences regarding how virtues and character manifest in the life of a soldier and officer.

Motivations for Joining the Army/Becoming an Officer

The reasons for becoming an Army officer given by interviewees varied and included a range of factors. Some officers mentioned having had a positive experience being in the Officer Training Corps (OTC), Combined Cadet Force (CCF) or Air Training Corps (ATC). Other reasons and motivations included having an interest in, and liking for, general participation in outdoor activities, including being a member of the Scouts and participating in sport; enjoying a challenge and believing that a career in the Army would provide this; wanting to lead people and make an impact; and incentives such as the provision of housing, disposable income and accruing a good pension.

DOI: 10.4324/9781003331209-5

Indeed, and more so than some of the other professions examined in Jubilee Centre studies such as nursing, medical practice and teaching, primarily officers cited non-moral reasons and/or motivations for joining the Army – as the following example illustrates:

> Because I didn't want to sit behind a desk all day which it turns out you've still got to do. I just wanted the variety. I wanted the opportunity to travel and the Army seemed to offer more than just sitting at a desk and managing.
>
> (CWC, female)

Being from a military family was a fairly common reason also cited by the officers in the interviews for their joining the Army, as the following two examples highlight:

> It [joining the Army] was something I was always going to do. My dad, my granddad, my great granddad, his dad, my uncle, my other uncle, my godfather, my other godfather – everyone has been in the army. No-one pushed me to do it, I just done that as well. It was just always something that it's not something I really thought about, I just did it.
>
> (JOTAC, male)

> My Dad was in the Army for 23 years. I was always exposed to the Army kind of lifestyle from a young age. I didn't really know any different. I don't think there's anything better than seeing the positive change in people as a result of your influence and I think that nothing really epitomises that more than being an Army Officer or a leader. The responsibility that you have and the responsibility to [other soldiers] that you have is phenomenal and the impact that you can make is quite big and it's something I've always wanted to do.
>
> (JOTAC, male)

In this latter extract, a real sense is gained of how non-moral motivations (in this case, exposure to the Army from a young age) can sit alongside moral motivations (i.e. making a positive change). A further example of the range of motivations sitting within one officer is also provided in the following explanation by another interviewee:

> I think I, probably deep down I was quite excited by the idea when I was younger but my parents weren't very keen, and I was in a job interview at the end of university and I had a bit of an epiphany and I thought I think this looks like fun, this is something I'd enjoy. I think I like the public service bit of it as well.
>
> (CWC, female)

Here, the officer explicitly references the importance of public service and in doing so implicitly connects with the culture and ethos of the Army in character terms.

Transferring Army Values and Standards between Professional and Personal Lives

Across the interviews there was general support for the idea that military values *do* transfer across professional and personal lives. As mentioned in Chapter 2, the Army Values and Standards of the British Ministry of Defence (2022) posit that a values, or virtue, approach is preferable to a strictly rule-based or deontological approach and that these values and standards translate to the personal lives of their members. Those interviewees who expressed the idea that Army Values and Standards *do* transfer across professional and personal lives typically offered two explanations as to why this was the case. The first set of explanations focused on the inherent connection between professional and personal values and standards, often suggesting that it was not possible to separate these domains. A second explanation concentrated not on the impossibility of a sharp distinction between professional and personal values in the lives of officers, but rather on the professional issues that would manifest if differences were at play. This second view was the more commonly expressed. Instances included the following:

> Yes I think it [transfer] absolutely should do [*sic*] especially because . . . everyone knows what everyone does and more so now than anything else. So you can't do, you know, one thing here and shut the door on your office and go 'yeah, see you later' and have two weeks doing whatever on holiday so, yes is my initial thought.
>
> (JOTAC, female)

A more detailed variation of this view was provided by another officer:

> I think it is difficult [to transfer into personal life] and some would argue 'why does the Army have a say over your personal life when your personal life is up to you?', which is true but you have to trust the people that you are working with. I think things such as affairs is [*sic*] a very easy one. I think it [*sic*] undermines trust, makes working relationships very difficult . . . when the rumour mill spreads very quickly and other people start finding out as well, which inevitably they do, I think it makes people question their [the soldier having the affair] ability as an officer to do their job and how can you do that when you haven't behaved correctly.
>
> (CWC, female)

Beyond this general view, there were interviewees who offered more nuanced accounts. One officer, for instance, who did think that transferral was possible and important gave the following explanation:

> I think it's [transfer] entirely possible and I wouldn't say it's unreason-
> able at all. The Army doesn't teach you to be a god at the end of the day.
> It teaches you to uphold a sense of morals and values which I think are a
> very good way to live your life . . . The Army's Values and Standards is
> just a good way to live your life I think it's a good character to be if you
> uphold those.
>
> (JOTAC, male)

When pushed as to whether there were difficulties in enacting this character, the officer suggested:

> There are always difficulties. It's not easy to be the best person you could
> be, it's hard and you're constantly striving to be that real good bloke, you
> know. Whether that's honesty or committing to everything, it's not easy
> but if you strive to achieve all of those then you're heading in the right
> direction if you ask me.
>
> (JOTAC, male)

Some other officers offered an alternative view of the transferral of values from professional to personal lives. One officer, for example, not only sug- gested that there should be a separation between professional and personal values but offered an explanation as to why this was important:

> I don't think it [transfer] can be – a lot of people would probably say 100%,
> but I don't think it can be. I think people need to be able to relax and be out-
> side of the army when they're on their leave or in their evenings. And they
> need an amount of privacy so that they can have their own lives. One day
> they're going to leave the army; it's important that they're not regimented.
>
> (CWC, male)

A few officers implied that the transferal of the Army Standards and Values was an ideal but one which was not always attainable or attained within the Army, as in this instance:

> Definitely, yeah, but I think that it's naive to suggest that we live by those
> all the time. I think in general, the officers, I mean, we are quite hypocriti-
> cal sometimes but in general I think we are better than soldiers, clearly
> there are lapses.
>
> (JOTAC, Male)

The findings and interview extracts presented in this section demonstrate how the values and standards of the army, including those that connect most readily with character and virtues, associate not just with soldiers' professional life but with their personal lives too. With regard to the transfer between the professional and personal, it is clear that the process is one which is not altogether straightforward and which requires ongoing effort and reflection.

What Personal Qualities and Character Strengths an Ideal Officer of Their Own Rank Might Have

Officers' views on the personal qualities and character strengths of an ideal officer of their own rank drew a range of responses. More general qualities cited ranged from being professionally competent, having enthusiasm, working hard, being charismatic and being dedicated. In addition, and importantly given the focus of this book, more specific strengths of character were also identified by interviewees. These more specific virtues included humility, honesty and selfless commitment. Interestingly, other than honesty, none of the virtues or character strengths the interviewees discussed explicitly in the interviews match what the officers rated as their top six overall virtues or character strengths (discussed in Chapter 3).

However, and in addition, what is most interesting in the responses to this question – as noted in the original report – is the way that some officers tended to focus on a range of qualities that, when taken together, comprise a well-rounded ideal officer. Often, as in the following two extracts, these collections of qualities were elucidated in terms of a general comprehensiveness in intention and action:

> Thoroughness, and by that I mean sort of being pretty thorough in your job and going out of your way to do things to make the soldiers' lives sort of better, be that getting them on courses or getting them extra training or whatever it happens to be. Those are the things that soldiers remember and actually respond to, to show that you're actually taking an interest in them and their careers and not actually just sort of doing the officer thing and not turning up or not showing up and not caring about it, if you're actually enthusiastic . . . they respond to that pretty well.
>
> (JOTAC, male)

> The soldiers they're loyal and they work hard and the best ones are the ones who even when the chips are down and things are really tough yet still have that quality that they will just push through and they will just get on with it and they know that at the end of the day it's their job and it's what they signed up to do. . . . Soldiers who work hard are outstanding and who consistently work hard regardless of what kind of condition we're in.
>
> (JOTAC, male)

A further theme of note in the interviews, connected with the qualities and character strengths an ideal officer of their own rank might have, related to role modelling. A number of the officers reported that learning from role models was important in the Army. For example:

> [Modelling] . . . Yeah, yeah, absolutely, I think we're constantly doing that. You see a guy presents to you, you work for a guy, you pick up on his habits and you see straight away like, what works well, how he approaches problems.
>
> (CWC, male)

The following officer spoke, for instance, about an officer they admired and had learned from:

> When I arrived there, he was certainly a role model. His attention to detail, for example, his kind of professional competence really kind of stood him out, and that was consistent in everything that he did. . . . On top of that, he had a very, very busy battalion with a lot of discipline problems. You know, he still managed to tick all the other boxes. He attended everything. The soldiers could see him at every PT session, despite – you know, no matter how much of a workload he had, he was there, doing whatever they were doing. . . . So . . . yeah, that's someone that I would admire.
>
> (CWC, male)

Once again, in this extract it is notable that rather than focusing on specific, individual qualities of character, the officer concentrates instead on a more connected sense of consistency and reliability in character.

Also identifiable in the officers' reflections on role modelling was the idea that rather than having a singular role model (as in the case above), officers drew from a range of role models, often drawing different learning from different colleagues:

> There isn't one person that I look at and go, 'I'm going to follow his example', but you do see people and think, you know, 'I need to be more like that'. There's definitely plenty of role models around that you take elements of.
>
> (CWC, female)

> I don't think it is necessarily just down to officers though. I would say I don't think it has to be . . . I think when you are going through training you have your staff sergeants/colour sergeants as well . . . I don't think necessarily it needs to be another officer. It could be maybe a senior NCO. I think they could also have the traits that you would admire.
>
> (CWC, female)

Finally, it is also relevant to note that some officers also spoke of learning from role models in relation to what might generally be considered as non-moral qualities. While non-moral in this sense, such qualities were clearly deemed important by the interviewees. When asked about role models in the Army, one officer gave the following response:

> An officer who is pretty organised, is aware of what's happening, when, and can see the important timelines and just understand really . . . because when you're doing timings for the military there's a lot going on. . . . There's just all sorts of activities, events, exercises, it's just a case of being able to actually pinpoint the important ones that are relevant. . . . The better officers are very strict with their timelines and know exactly when they need to get the notes in, when [they] need to have written their reports etc.
>
> (CWC, female)

Which Personal Qualities or Strengths Are Most Important to them in Their Current Role as an Army Officer

Officers were asked about the personal qualities or strengths that they perceived to be most important to them in their current role as an Army officer. As with the ideal qualities or character strengths above, responses ranged from more general qualities – such as competence, being responsive to feedback, motivation, humour, respect for others and teamwork – and more specific character strengths or virtues – such as integrity, loyalty, courage, leadership and honesty.

Given their importance to the Army, in the interviews officers were questioned about *self-discipline* and *selfless commitment* if these values were not mentioned spontaneously by the interviewee. Officers presented their relationship with the values of self-discipline and selfless commitment in different ways, suggesting that, while these stand out as general principles to guide conduct, officers have their own more specific interpretations and ways of enacting them. Some officers suggested that although they were aware of the importance of self-discipline and selfless commitment, as officers they were 'not using them all the time' (JOTAC, male).

For other officers, selfless commitment in particular was a core part of being a soldier. These officers reported how selfless commitment and self-discipline were qualities they strived for in their work, as the following two extracts illustrate:

> For me personally I'd say it's [selfless commitment] hugely important. I think it's massively important. If you are not signed up to the Army way of life and if it is not for you and you have qualms or there are things that you don't agree with, there are things that you don't want to do, then you're showing a lack of selfless commitment if you ask me. At the end of

the day you sign up to the Army . . . you're told you deal with it. The needs of the Army come first . . . family come second. It's becoming harder in this day and age. I would say that it's slipping in the sense that people now feel like it's becoming more of a job as opposed to an actual vocation and a way of life that you sign up for. . . . It's not just turning up for work in the morning and leaving at 5 o'clock in the evening. It's a whole, it's a community, it's a family, and your commitment to that I think is paramount at the end of the day.

(JOTAC, male)

Self-discipline [is vital in] all areas of Army life . . . I think it's hugely important . . . it's having that discipline to walk away if things are going really wrong or deal with a situation or know that something is going wrong and have an impact on it. There is a line, there is always a line, and discipline, whether that's when you're out or whether you're in barracks, and how you act is I think it vitally important.

(JOTAC, male)

Other officers were more hesitant about placing labels on particular forms of conduct, as in the following instance in which the officer alludes to the fact that while actions may comprise selfless commitment, the actual term may not always be front and centre in the day-to-day life of soldiers:

Selfless commitment, it sounds a bit . . . I like to feel like I have done my utmost for the guys as much as I can whether it is in terms of posting or making sure their reports are done, giving them leave if they need leave. I like to feel that I have done as much as I can. Sometimes you do as much as you can and they still throw it back in your face. Yes – ill always try and work late and try and be as on top of things as I can be because ultimately that's for their benefit.

(CWC, female)

In several of the interviews it was clear that the officers were thinking and reflecting deeply about how they practised self-discipline and selfless commitment, and the challenges involved in doing so. In these cases, the officers considered the difference between *professing* selfless commitment and what it meant to *enact* selfless commitment, as well as its delimits – including where selfless commitment could be excessive, as in the following:

I'm quite committed to my job. Like, probably over-committed. So I'm going through a divorce at the moment. My wife has stated that that's part of the reason; said, you know, 'You won't let work go. You'll sit and work till 10 at night.' And she gets that. You know, she was like, 'It's just 'cause you want to do your job well. You don't want to see it fail.' But I think a lot of army officers have got that. It's just making a point where to draw the line. Yeah, I'd say I'm quite committed to the job, even to my own loss, at times.

(CWC, male)

Similarly, these officers also considered in the interview the extent to which self-discipline is a somewhat ambiguous quality, open to interpretation and flexibility – a fact that can bring positives and negatives.

Pressures and Barriers that Constrain Ethical Action

When asked about routine professional challenges faced, army officers again cited a range of contextual and organisational factors that were a concern. These are perhaps readily apparent for anyone who has served in an army, and while some may seem somewhat mundane or just part and parcel of being in a large and complex modern organisation, their import should not be under-estimated given the attention paid to organisational constraints that impact negatively on character within the general literature on professional ethics (as outlined in Chapter 1). Such concerns incorporated uncertainty, welfare issues (such as accommodation) and handling the constant changes to life involved. A number of officers spoke of the challenges of handling relationships within the Army, including communicating with other soldiers of different ranks.

When officers were asked about those pressures or barriers that impede on their ability to act ethically in their professional role, once again a range of factors were cited across the interviews. One officer, for example, mentioned the pressure of facing 'repercussions . . . if you do the right thing not everyone likes it . . . you then have to work with people after as well which is difficult' (CWC, female). This response speaks to the importance of ongoing cultures, norms and practices within professional settings such as the Army and how these can place real pressure on professionals to enact their moral identity. Another case in point cited by several officers was accountability, which, as the following response suggests, was distinguished from moral or personal accountability:

> The accountability factor does come in, but that is not in terms of personal like, 'yeah, I did that, I did that for these reasons, and these reasons determined my actions' – absolutely, you know, we all need accountability. Sometimes it's the, you know, just the work that generates, the paperwork that you have to go through?

(CWC, male)

When asked about pressures and barriers, the following officer contemplated how the culture of the Army influenced moral decision-making by emphasising the importance of how soldiers are perceived by other soldiers and positioning this as an obstacle to moral conduct:

> I suppose you can weight it up on, 'Will anyone ever know?' I think peo-ple, a lot of the time, they care about what other people think about them, so they think, 'Well, if no-one ever knows about this, no-one will think less of me, so, you know . . .' and they balance that against the difficulty of doing the right thing. 'Cause if the right thing's easy, and the easiest thing to do, then people, of course, are always going to do it. No-one does the wrong thing to make life harder for themselves. They're doing the wrong thing to make life easier or, you know, to make money, if it was theft, for example, or make their own lives comfortable . . . . I think, a lot of the time, it's them weighing up, 'What will people think about me?' And, really, that should be removed. It should be kind of . . . if no-one knows, you know, I should still do the right thing here.
>
> (CWC, male)

Another officer spoke in these terms about the organisational systems that place pressure on the moral dimensions of their professional role:

> I think the way the organisation is, it's a very hierarchical organisation and sometimes when someone above you says 'Yeah, just do this' and you know it's kind of a little bit grey, then you could be tempted to say 'I'll still do that' even though you know it's a bit, or you're tempted or forced to do that even though, 'cause a person in authority has said 'Yeah you can do it'. . . . Everyone wants to like deliver the end results and when that competes with the moral thing then sometimes there's which is more important?
>
> (JOTAC, male)

At the end of this extract, a real picture is gained of the different pressures at play and the need for officers to make important choices and decisions. This brings into sharp focus the importance of *phronesis* and professional judgement. Though the terms *phronesis* or practical wisdom were not used explicitly, some other officers made reference to the challenges of arriving at a clear and balanced decision about correct action given the competing priori-ties involved:

> So do the right thing, yeah great. OK we all get told to do the right thing, but actually putting that into practice . . . you get some quite interesting situations where actually you're trying to balance what's best. Actually some of the hardest ones are when you're trying to judge what's best for your recruits and what [is best for] squadron headquarters. . . . You sit in

the middle with your troop and then you've got the squadron above you and you're thinking I know them, this is how I want to deal with it, yes of course please give me your advice but I usually try and go in and go 'this is the problem, this is what I propose'.

(JOTAC, female)

Conclusion

The analysis in this chapter, including the extracts from the officers interviewed, sheds some light on how these soldiers in the British Army perceive and experience values, virtues and character. The data suggests that while moral motivations were at times in play in terms of why the respondents joined the Army and wished to become officers (including taking responsibility and impacting positively on others), these were often implicit and/or stood alongside non-moral factors. Though often noting the complexities involved, officers largely agreed that Army values do transfer between professional and private life, with some holding the interesting view that the values of the British Army offer a good basis for personal conduct. When asked about specific qualities, a range of general qualities and more specific values were cited across the interviews. Alongside these, a notable aspect of the responses was how a number of the officers spoke of the 'well-rounded' officer who possesses a range of virtues and exhibits 'reliability of character'. In addition, when asked, the officers all spoke of the importance of self-commitment and self-discipline, though there was a sense in the data that they each had their own way of interpreting and defining what these terms meant in practice. Finally, when speaking of the pressures and barriers at play, three stood out that are of particular interest given the wider literature and studies on professional ethics – the significance attached to standing in relation to others, the constraints of organisational systems and structures, and the challenge of discerning the right course of action in complex situations and given competing demands. In the final chapter of this book, to which we now move, some overall conclusions are presented before a number of recommendations are offered.

Conclusions, Recommendations and Further Research

Introduction

Being a member of a 'profession of arms' necessarily involves soldiers in a range of ethical situations and, when operating properly, in considering how the soldier's personal conduct relates to their professional role and activity as a soldier and, indeed, how one's conduct stands in relation to the values, culture and ethos of the particular professional army in which the soldier serves. As others have argued, and as we have contended in this book, being a soldier involves a range of virtues. These virtues include courage, humility, honesty, loyalty and self-discipline. While commitment to these virtues at a general level is not always contested, when specific situations are involved – as with the dilemmas used within the AICM developed in the *Soldiers of Character* study – the meaning, enactment and prioritisation of these virtues become more complex, for instance where loyalty to one's fellow soldier's is in conflict with honesty. In addition, and as we have pointed to in places within this book, the changing nature of modern warfare presents new challenges for military ethics and for what it means, ethically speaking, to be a good soldier.

In Chapters 3 and 4 of this book, we presented and analysed a range of empirical data to present a picture of the soldiers' ethical reasoning in relation to the four dilemmas in the AICM, including how this reasoning stood in relation to the best and worst action and justification decisions determined by the expert panel, the soldiers' self-reports of virtues important to who they are and interview data to provide a picture of how the officers perceived and experienced values, virtues and character in the British Army. In this final chapter we set out in short form some of the key findings of the *Soldiers of Character* study presented in the book, before offering some recommendations based on the preceding analysis. Finally, we advance several areas in this field that would benefit from further research.

DOI: 10.4324/9781003331209-6

Overall Findings

Key findings of the *Soldiers of Character* study and the additional analysis given in these pages can be summarised as follows:

- Officers' responses to the dilemmas were close to the judgements of the expert panel in regard to the actions and the justifications selected. The positive overall scores for the AICM (M = 0.65) suggest that the officers were appropriately applying Army values in their responses to the dilemmas and also that their ethical reasoning aligned well with Army standards of excellence.
- In the AICM responses, officers were closer to the judgements of the expert panel when selecting *actions* than when selecting *justifications*.
- Officers' responses to the VIA-IS-EI suggested that certain virtues – namely teamwork, honesty, curiosity, fairness, perseverance and leadership – were viewed by soldiers as the most important to who they are in the world. These are well-aligned with the values of the British Army.
- Importantly, higher self-reported averages for strengths such as judgement, honesty, bravery and leadership corresponded to higher total AICM scores.
- In the interviews, officers suggested that although ethical factors were at times involved in motivations for joining the Army, these were often implicit and/or stood alongside non-moral factors.
- Though clearly involving a number of complexities, the officers in the study largely considered Army values to transfer between professional and private life, with some holding the interesting view that the values of the British Army offer a good basis for personal conduct.
- Though specific qualities were highlighted by the officers as important for being a good soldier, a notable aspect of the interview responses was that a number of the officers spoke of the 'well-rounded' officer who possesses a range of virtues and exhibits a 'reliability of character'.
- Officers cited a number of pressures and barriers at play that could constrain the enactment of good character, with three of particular interest standing out – the significance attached to standing in relation to others, the constraints of organisational systems and structures, and the challenge of discerning the right course of action in complex situations and given competing demands.

Further Research

The analysis of data presented in the *Soldiers of Character* project's final report, and that set out in this book, points to some key areas that would benefit from further research. These areas include the following:

- Further use of the AICM as a measure that specifically examines, and provides insights about, the ethical reasoning of soldiers at different ranks than that involved in the *Soldiers of Character* study

- Further interrogation and understanding of whether, and if so how, certain virtues – such as judgement, honesty, bravery and leadership – corresponded to higher total AICM scores as suggested in the data collected in the *Soldiers of Character* study
- Further investigation – theoretical and empirical – of the place of character and virtues in the context of the changing nature of modern armies and warfare, including those stemming from developments in Artificial Intelligence
- Further qualitative studies that examine how soldiers of different ranks adjudicate between virtues when these conflict in practice
- Further clarification, empirical research about and practical enactment of educational endeavours at different levels of the professional development of soldiers that attend to the ethical dimensions of being a soldier, including the character and virtues necessary for being an ethically good soldier – in particular, the cultivation of professional *phronesis*.

Appendices

Appendix 1

Dilemma 1 – Captain Metcalf

Capt Metcalf is the OC[1] of a RMP[2] Company in Somalia. Metcalf's job as part of the African Union Mission in Somalia, operating under UNSCR[3] 2182 (2014), is to organise humanitarian assistance to the Somali people and make southern Somalia safer. Armed militia and rival groups still exist in Somalia. This makes it a hostile place for giving humanitarian aid and economic help. Capt Metcalf discovered in command briefs that Somali militia often use innocent civilians as human shields or decoys – these can be women, children or old people. This happened two days ago for Capt Metcalf's unit when two of Metcalf's soldiers were ambushed and killed while they were checking out an incident reported by a woman. Today, Capt Metcalf and three soldiers are driving in a Jackal vehicle (armed with a GPMG)[4] through Mogadishu. They are supported by another vehicle of the same kind with two more soldiers. They are en route to a checkpoint on a non-time sensitive resupply tasking. Before reaching the checkpoint they discover an injured male Somali covered in blood from his belly to his feet. He is clearly wounded and is surrounded by a large unpredictable crowd of locals. One of the three soldiers tells Metcalf that he knows the injured man, and Metcalf realises that the man recently provided high-value information that led to the capture of several militia. Metcalf knows of the mandate to protect civilians and the requirement to take all reasonable measures necessary for the collection and care of those injured, whoever they are. No one in the vehicle knows how the man was injured, but the present location is a high-risk area for ambush. The British RMP patrol is armed only with their personal weapons and a GPMG – there's no way they could fight off an ambush. The bloodied Somali is sure to die if he doesn't get immediate medical help. What should Metcalf do?

For this dilemma, respondents were given nine actions that Captain Metcalfe should take to deal with the situation and asked to rate these on a scale of 1–5 (1 = I strongly believe this is a GOOD choice; 2 = I believe this is a

GOOD choice; 3 = I am not sure; 4 = I believe this is a BAD choice; and 5 = I strongly believe this is a BAD choice) before selecting the best three choices and the worst three choices. The nine actions were:

1. Bypass the injured Somali and continue the assigned mission
2. Stop, provide medical assistance, and call and wait for medical support
3. Extract from the area immediately and go back to base
4. Order the soldiers to move in, lift and place the injured Somali into their vehicle and transport him to the nearest aid station
5. Radio back to headquarters to request guidance on how to handle the situation
6. Continue the mission, but leave two soldiers with the injured Somali to provide security and assistance until a requested medical ambulance arrives
7. Retrieve the Somali, provide medical assistance and transport him to the nearest aid station after completing her mission
8. Secure the area and await local assistance
9. Bypass the injured Somali but call in the known location and request medical support with armoured escort

Respondents were then asked to judge a set of reasons that Captain Metcalfe might be thinking about while making the decision, rating these on a scale of 1–5 (1 = I strongly believe this is important; 2 = I believe this is important; 3 = I am not sure; 4 – I believe this is not important; 5 = I strongly believe this is not important) before selecting the three most important reasons and the three least important reasons. The reasons provided were:

1. A Somali's life is not worth risking the lives of British soldiers
2. Stopping to help could affect our ability to accomplish our assigned mission
3. We didn't injure the Somali, so let his people take care of him
4. The injured Somali has helped us, so we should help him
5. It will weigh on my conscience if I leave an injured man I could have helped
6. I shouldn't risk six lives to save one
7. The mission doesn't call for me to stop and help a Somali
8. My soldiers could get hurt or killed
9. Even if we stop and help, it looks like the Somali will die
10. I should attempt to accomplish both missions: help a Somali that has helped us and complete the resupply task
11. If we helped and one of my soldiers were injured or killed, my unit would condemn my actions

12. I must do what I am ordered to do
13. We are explicitly mandated to collect and care for the wounded
14. Leaving him to die might discourage other informants from helping us
15. I could get hurt or killed
16. The injured Somali could have additional information that could lead to the capture of more militia

Dilemma 2 – Major Smith

During major combat operations, Major Tim Smith, OC 'A' Company, has been ordered to withdraw to a pre-planned defensive position due to heavy enemy activity. 'A' Company is all ready to go when a warning order comes in. This gives Smith new orders to move to another location and prepare an ambush on an enemy platoon who are expected to pass through that location. Smith is told it is essential that 'A' Company successfully executes the ambush to enable the battle group to establish itself in the defensive location. No other company can reach the ambush position in time. Just as everyone prepares to follow the order, a section recce patrol returns to the company location in some distress. They report to Maj Smith that they saw Corporal Taylor and Private Edmonds, a second recce patrol, being captured by a squad of enemy soldiers. They followed the enemy squad into dense vegetation and captured one enemy fighter at the rear. This prisoner said that his squad was taking the British soldiers to what he called 'a safe house', but he refused to say where it is. The Sergeant Major says: 'We have to get them back! Remember Sgt Wright and Cpl Field – when this happened to them they were burnt and their bodies dragged through the streets. Taylor and Edmunds will be lucky if they are just shot! Give me 5 minutes with the prisoner the patrol brought back and I will get the location of our guy. He will beg to talk to me. If close, we can snatch them back quickly'. A check of the distance to the ambush position convinces Smith that he has time to rescue the men. Smith needs to get the location from the enemy prisoner. Taylor and Edmonds will probably be tortured and brutally killed unless quickly rescued. Nearest military interrogators are three hours from their location. What should Major Smith do?

For this dilemma, respondents were given nine actions that Captain Metcalfe should take to deal with the situation and asked to rank these on a scale of 1–5 (1 = I strongly believe this is a GOOD choice; 2 = I believe this is a GOOD choice; 3 = I am not sure; 4 = I believe this is a BAD choice; and 5 = I strongly believe this is a BAD choice) before selecting the best three choices and the worst three choices. The nine actions were:

1. Let the Sergeant Major question the prisoner alone
2. Let only a trained interrogator question the prisoner
3. Use any means available to get the location of the enemy 'safe house'

4. Turn a blind eye to the Sergeant Major's intent
5. Force the prisoner to guide a rescue party to the enemy 'safe house'
6. Report the prisoner and move the company immediately to its ambush position
7. Aggressively interrogate the prisoner personally to get the location of the captured soldiers, but afterwards, Smith should turn himself over to the superiors to take responsibility for his actions
8. Abandon the mission, prioritise the soldiers
9. Monitor the Sergeant Major's interrogation of the prisoner to ensure proper procedures are followed

Respondents were then asked to judge a set of reasons that Major Smith might be thinking about while making the decision, rating these on a scale of 1–5 (1 = I strongly believe this is important; 2 = I believe this is important; 3 = I am not sure; 4 – I believe this is not important; 5 = I strongly believe this is not important) before selecting the three most important reasons and the three least important reasons. The reasons provided were:

1. Whatever the Sergeant Major does to get the information won't be as bad as what the enemy will do to our soldiers if we don't rescue them
2. I am responsible for the actions of the Sergeant Major
3. We should not abandon our soldiers
4. Two soldiers' lives are much more important than the lawful treatment of one prisoner
5. No one in the company would corroborate that I tortured the prisoner
6. Our mistreatment of this prisoner could lead to future mistreatment of other captured British soldiers
7. The unit might overlook any extreme measures taken against prisoners if we obtained vital information
8. Established procedures need to be followed when handling prisoners
9. The prisoner has rights that must be protected
10. The only way to get results from this enemy is to use the same tactics they use
11. If we don't get the soldiers back, my soldiers will lose confidence in me
12. The death of these two soldiers will negatively affect the morale of my unit
13. I might be relieved of command if higher HQs find out we mistreated prisoners
14. I will risk the reputation of the Regiment and British Army if I allow mistreatment of the prisoner
15. Because of their actions, the enemy has lost their rights to be treated as captured personnel
16. Torture is wrong but morally permissible to save lives

Dilemma 3 – Lieutenant Colonel Milgram

Lt Col Milgram has just been to a memorial service for one of the Unit's best junior officers – Captain Ben Richards. Richards' was killed from an IED,[5] and Milgram put him into the ambulance. This death brought a close unit even closer together. They are under constant fire from insurgents, and the situation had deteriorated into 'armed conflict'. Milgram respects all of the soldiers who are wearing themselves out to succeed in tasks given to the Battalion, and Milgram told Unit leaders that a top mission is to protect their soldiers. It was the holy month of Ramadan, which coincided with an increase in attacks and British combat deaths. Milgram's unit, because of their aggressiveness and positive results, was given the mission in the insurgency-ravaged Anbar province to bring the Iraqi city of Samara back under control. Milgram's unit was authorised a range of aggressive tactics. A curfew was in force throughout Iraq to help separate insurgents from ordinary Iraqis. Any person caught after curfew would be detained. It was just so hard to separate the good from the bad. On the day of Capt Richard's memorial service, one of Milgram's company commanders informed him that a squad using nonlethal force might have accidently killed two Iraqi civilians. Lt Col Milgram knew about several nonlethal tactics and had condoned them. Lt Col Milgram's superiors had also condoned these nonlethal tactics. However, Milgram never authorised the tactic that was used on the two possible dead Iraqis, which involved throwing them into the river. The company commander told Milgram that their soldiers had been throwing Iraqis in water as a deterrent for curfew violations and considered it within the scope of nonlethal tactics. Milgram knew that a criminal investigation would occur and that dedicated young soldiers may well be arrested. What should Milgram do?

For this dilemma, respondents were given eight actions that Lt Col Milgram should take to deal with the situation and asked to rank these on a scale of 1–5 (1 = I strongly believe this is a GOOD choice; 2 = I believe this is a GOOD choice; 3 = I am not sure; 4 = I believe this is a BAD choice; and 5 = I strongly believe this is a BAD choice) before selecting the best three choices and the worst three choices. The eight actions were:

1. Instruct the soldiers to tell the investigators everything except about the water
2. Explain to his superiors everything that had happened
3. Take responsibility for the soldiers' actions by telling his superiors that he had authorised the throwing of curfew violators into rivers
4. Instruct the soldiers to be honest in their statement to the investigators
5. Tell superiors that Milgram was unaware of the soldiers' tactics and had not condoned them
6. Tell superiors that the soldiers did not throw the Iraqi civilians in the river
7. Tell the soldiers to deny the allegations of throwing Iraqis into rivers
8. Defend the soldiers based on the stress they have been under and the vagueness of the ruling about nonlethal force – their intent was not to drown

Respondents were then asked to judge a set of reasons that Lt Col Milgram might be thinking about while making the decision, rating these on a scale of 1–5 (1 = I strongly believe this is important; 2 = I believe this is important; 3 = I am not sure; 4 – I believe this is not important; 5 = I strongly believe this is not important) before selecting the three most important reasons and the three least important reasons. The reasons provided were:

1. After all that my men have done for me, I can't let them be court-martialled
2. Allowing this incident to become public knowledge will only increase anti-British feeling among the Iraqi people
3. Taking responsibility for our actions is the right thing to do
4. If I cover up this incident, I will be placing my career in jeopardy
5. Our superiors, who have not experienced the environment we are operating in, should never have placed my soldiers in this predicament where they felt forced to use this type of tactic
6. My soldiers and I aren't really lying if we just fail to say exactly what happened
7. It is my duty to put the welfare of my soldiers before the truth
8. My superiors' vague guidance on nonlethal tactic has caused this problem
9. Lying to my superiors goes against my own values
10. If I tell my superiors what happened, I will be placing my career in jeopardy
11. If I don't protect my soldiers and myself, higher HQs will just use us as scapegoats

Dilemma 4 – Lieutenant Jacobs

Things couldn't be going any better for Lt Jacobs. After months of hard training, he had earned his wings after passing P Company. His best friend, Lt Drake, who won the 'Sword of Honour' at Sandhurst, had really helped him prepare for P Coy. Drake, considered the best lieutenant in the battalion, had used most of his free time to make sure Jacobs was ready for the challenges of P Coy. Jacobs and Drake had become inseparable since Sandhurst. Drake had always done well and had helped Jacobs succeed as well. The leadership in the Battalion, especially the CO,[6] really liked Drake and saw in him tremendous potential. Soon after earning his wings, Jacobs and Drake celebrated down town in their usual pub. Quite soon, Drake meets a pretty woman and her friend. The four spend the next few hours talking and drinking. Jacobs can tell Drake really likes the women he is with. However, when it is time to leave, the women tell them that they are junior non-commissioned soldiers newly posted to the Battalion. Jacobs immediately tells them that they are officers and that it is against regulations to have a personal relationship with them. Drake seems annoyed at Jacobs' comments and whispers something in the women's ear before she leaves. Jacobs reminds Drake about the briefing they received a month ago warning about having any kind of personal relationship with other ranks in the Battalion. Drake tells Jacobs that he doesn't have to worry about him and that he won't be seeing her again. However, a week later, as Jacobs is driving out of camp, he sees Drake and the young women coming out of a shop holding hands and later in the week he sees them in a passionate embrace. Jacobs also learns from a reliable source that Drakes' relationship with the women is sexual. That night Jacobs calls Drake to ask what is going on with the women. Drake says he is dating the soldier, plans to continue to do so, and that he knows that Jacobs will keep quiet. What should Jacobs do?

For this dilemma, respondents were given nine actions that Lt Jacobs should take to deal with the situation and asked to rank these on a scale of 1–5 (1 = I strongly believe this is a GOOD choice; 2 = I believe this is a GOOD choice; 3 = I am not sure; 4 = I believe this is a BAD choice; and 5 = I strongly believe this is a BAD choice) before selecting the best three choices and the worst three choices. The ten actions were:

1. Tell Lt Drake that he won't cover for him when it comes to inappropriate relations between officer and other ranks
2. Do nothing and let Lt Drake continue the relationship with the soldier
3. Tell the chain of command about Lt Drake's improper relationship
4. Tell the female soldier to stop the relationship
5. Warn Lt Drake that you will inform the chain of command unless he ends the relationship
6. Tell another peer about Lt Drake's relationship and see if he informs
7. Express his resentment to Lt Drake for putting him in this dilemma, but do not tell the chain of command
8. Send an anonymous note to the Commanding Officer about Lt Drake's prohibited relationship
9. Tell Lt Drake to inform the chain of command about the relationship or he will have to
10. If the relationship really is serious, Drake should be persuaded to tell the chain of command so that he can find a way to continue it legitimately

Respondents were then asked to judge a set of reasons that Lt Jacobs might be thinking about while making the decision, rating these on a scale of 1–5 (1 = I strongly believe this is important; 2 = I believe this is important; 3 = I am not sure; 4 – I believe this is not important; 5 = I strongly believe this is not important) before selecting the three most important reasons and the three least important reasons. The reasons provided were:

1. I should be loyal to my best friend
2. My peers will resent me if I inform the chain of command about Lt Drake

3. Lt Drake would 'turn-a-blind-eye' if it were me, so I should too
4. Lt Drake's relationship could get him in trouble
5. Lt Drake lied to me about the relationship
6. By doing what is right, I will gain the respect of my leaders
7. I could get in trouble if Lt Drake gets caught and the leadership finds out that I knew about the relationship
8. The Army fraternisation policy is pointless anyway
9. Lt Drake's actions compromise authority discipline and the morale of our unit
10. It is my responsibility to take corrective action against any soldier not following Army regulations
11. Everybody else seems to look the other way when soldiers and officers violate this policy
12. As a British officer, I am duty bound to tell the truth

Appendix 2 Demographic Categories

Variable	Category	Number	(%)
Seniority by rank and service	Cadet	76	31
	Lieutenant and junior Captains		
	– Captain with five or less years' service	93	38
	Senior Captains and Majors		
	– Captain with six or more years' service	73	30
Seniority by course	Royal Military Academy Sandhurst	76	31
	Junior Officer Tactical Awareness Course	85	35
	Captains Warfare Course	81	36
Gender	Male	190	78
	Female	52	21
Ethnicity	White British, Irish, Other White	234	97
	Indian, Chinese, Asian, Mixed	5	2
	Rather not say	3	1
Religion	Christianity	135	56
	Atheist	70	29
	Buddhism, Judaism, Hinduism	4	2
	Other	11	5
	Don't know	10	4
	Rather not say	9	4
	Did not answer	3	1
Future career intentions	Stay as long as I can	144	60
	Leave at end current contract	51	21
	Leaving	35	15
	Did not answer	12	5

(Continued)

Appendix 2 (Continued)

Variable	Category	Number	(%)
Type of Commission	No commission (officer cadet)	74	31
	Short Service Commission	88	36
	Intermediate Regular Commission (IRC)	51	21
	Regular Commission	21	9
	Did not answer	8	3
Education level	Below degree	39	16
	Degree	159	66
	Post degree	38	16
	Did not answer	6	3
Number of operational tours	None	138	57
	1 or 2	74	31
	More than 2	15	6
	Did not answer	15	6
Self-rating compared to peers	Better	34	14
	Mostly better	103	43
	About same	96	40
	Did not answer	9	4
Age	Age 30 and below	186	77
	Age 31 and above	53	22
	Did not answer	3	1
Corps/kind of service	Royal logistic Corps	28	12
	Adjutant General Corps	20	8
	Royal Engineers	20	8
	Royal Electrical and mechanical Engineers	15	6
	Infantry	65	27
	Royal Artillery	19	8
	Royal Military Police	10	4
	Royal Signals	14	6
	Medical/Veterinary	5	2
	Intelligence Corps	10	4
	Army Air Corps	10	4
	Not yet allocated – RMAS	26	11
Branch of service	Infantry/artillery	84	35
	Non-infantry/artillery	158	65

Notes

1 Officer Commanding.
2 Royal Military Police.
3 United Nations Security Council Resolution. UNSCR requires the force to prevent abuse/violence to women.
4 General Purpose Machine Gun.
5 Improvised explosive device.
6 Commanding Officer.

References

Annas, J. (2011) *Intelligent Virtue*. Oxford: Oxford University Press, pp. 8–15, 120–131.

Aristotle (1985) *Nicomachean Ethics*, trans. Irwin, T. Indianapolis: Hackett Publishing.

Armstrong, A.E. (2007) *Nursing ethics: a virtue-based approach*. Basingstoke: Palgrave Macmillan.

Arthur, J. and Earl, S. (2020) *Character in the Professions: How Virtue Informs Practice*. Birmingham: Jubilee Centre for Character and Virtues, University of Birmingham.

Arthur, J., Earl, S., Thompson, A., and Ward, J. (2019) 'The value of character-based judgment in the professional domain', *Journal of Business Ethics*. DOI:10.10007/s10551-019-04269-7.

Arthur, J., Kristjánsson, K., Thomas, H., Holdsworth, M., Badini Confolonieri, L. and Qiu, T. (2014) *Virtuous Character for the Practice of Law: Research Report*. Birmingham: Jubilee Centre for Character and Virtues, University of Birmingham.

Arthur, J., Kristjánsson, K., Thomas, H., Kotzee, B., Ignatowicz, A. and Qiu, T. (2015a) *Virtuous Medical Practice*. Jubile Centre for Character and Virtues.

Arthur, J., Kristjánsson, K., Thomas, H., Kotzee, B., Ignatowicz, A., and Qiu, T. (2015) *Virtuous Medical Practice: Research Report*. Birmingham: Jubilee Centre for Character and Virtues, University of Birmingham.

Arthur, J., Kristjánsson, K., Walker, D., Sanderse, W., and Jones, C. (2015) *Character Education in UK Schools*. Birmingham: Jubilee Centre for Character and Virtues, University of Birmingham.

Arthur, J. and Peterson, A. (2021) *Ethics and the Good Teacher*. London and New York: Routledge, pp. 56–61.

Arthur, J., Walker, D.I., and Thoma, S. (2018) *Soldiers of Character: Research Report*. Birmingham: Jubilee Centre for Character and Virtues, University of Birmingham. Available at www.jubileecentre.ac.uk/userfiles/jubileecentre/pdf/Research%20Reports/Soldiers_of_Character. Pdf

Australian Army. (2022) Our values and contract. Available at: www.army.gov.au/our-people/our-values-contract

Barnes, D.M. (2016) *The Ethics of Military Privatization: The US Armed Contractor Phenomenon*. London and New York: Oxford University Press.

Battistelli, F. (2000) 'The Postmodern Military: Conscription or Professionalism?', in Cohen, S.A. (ed), *Democratic Societies and Their Armed Forces: Israel in Comparative Context*. London and Portland, OR: Frank Cass Publishers, pp. 32–79.

Bayles, M.D. (1988) 'The Professions', in Callahan, J. C. (ed) *Ethical Issues in Professional Life*, New York: Oxford University Press, pp. 27–30.

Bebeau, M.J. and Thoma, S.J. (1994) 'The impact of a dental ethics curriculum on moral reasoning', *Journal of Dental Education*, 58 (9), pp. 684–692.

Bebeau, M.J. and Thoma, S.J. (1998a) 'Expert novice differences on a measure of intermediate level ethical concepts', Paper Presented at the *Annual Meeting of the American Association of Dental Schools*, Minneapolis, MN, Feb. 28.

Bebeau, M.J. and Thoma, S.J. (1998b) 'Designing and testing a measure of intermediate level ethical concepts', Paper Presented at the *Annual Meeting of the American Educational Research Association*, San Diego, CA, Apr. 13–17, pp. 1–19.

Bebeau, M.J. and Thoma, S.J. (1999) 'Intermediate concepts and the connection to moral education', *Educational Psychology Review*, 11, pp. 343–360.

Bellamy, A.J. and Dunne, T. (2016) 'R2P in Theory and Practice', in Bellamy, A.J. and Dunne, T. (eds.), *The Oxford Handbook of the Responsibility to Protect*. Oxford: Oxford University Press, pp. 3–18.

Bellamy, A.J. and Luck, E.C. (2018) *The Responsibility to Protect: From Promise to Practice*. Cambridge: Polity Press.

Bessant, J. (2009) 'Aristotle meets youth work: A case for virtue ethics', *Journal of Youth Studies*, 12 (4), pp. 423–438.

Bieri, M. and Dickow, M. (2014) 'Lethal autonomous weapons systems: future challenges', *Center for Security Studies Analyses*, 164, pp. 1–4. Available at: www.research-collection.ethz.ch/bitstream/handle/20.500.11850/91585/1/eth-46945-01.pdf

Biggar, N. (2013) *In Defence of War*. Oxford: Oxford University Press.

Blond, P., Antonacopoulou, E. and Pabst, A. (2015) *In Professions We Trust: Fostering Virtuous Practitioners in Teaching, Law and Medicine*. London: Respublica.

Boe, O. (2015) 'Building Resilience: The Role of Character Strengths in the Selection and Education of Military Leaders', *International Journal of Emergency Mental Health and Human Resilience*, 17 (4), pp. 714–716.

Boddens Hosang, J.F.R. (2020) *Rules of Engagement and the International Law of Military Operations*. Oxford: Oxford University Press.

Bohlin, K. (2005) *Teaching Character Education Through Literature: Awakening the Moral Imagination in Secondary Classrooms*. London: Routledge.

Bontemps-Hommen, C.M.M.L., Baart, A. and Vosman, F.T.H. (2019) 'Practical wisdom in complex medical practices: a critical proposal', *Medicine, Health-Care and Philosophy*, 22, pp. 95–105.

Brahnam, S. (2009) 'Building character for artificial conversational agents: ethos, ethics, believability, and credibility', *PsychNology Journal*, 7 (1), pp. 9–47. Available at: www.psychology.org/File/PNJ7(1)/PSYCHNOLOGY_JOURNAL_7_1_FULL.pdf#page=9

Braun, V. and Clarke, V. (2006) 'Using thematic analysis in psychology', *Qualitative Research in Psychology*, 3 (2), pp. 77–101.

British Army. (2022) *A soldier's values and standards*. Available at: www.army.mod.uk/who-we-are/our-people/a-soldiers-values-and-standards/

Brody, H. and Doukas, D. (2014) 'Professionalism: A framework to guide medical education', *Medical Education*, 48 (10), pp. 980–987.

Bunch, W.H. (2005) 'Changing moral judgement in divinity students', *Journal of Moral Education*, 34 (3), pp. 363–370.

Burk, J. (2002) 'Expertise, Jurisdiction, and the Legitimacy of the Military Profession," in Snider D.M. and Matthews, L.J. (eds.), *The Future of the Army Profession*. Boston: McGraw-Hill Primis Custom Publishing, pp. 21.

Burkhardt, T. (2017) *Just War and Human Rights: Fighting with Right Intention.* Albany: SUNY Press.

Caforio, G. (2006) *Handbook of the Sociology of the Military.* New York: Springer.

Campbell, C.P. (1995) 'Ethos: character and ethics in technical writing', *IEEE Transactions on Professional Communication,* 38 (3), pp. 132–138. Available at: https://ieeexplore.ieee.org/stamp/stamp.jsp?arnumber=406725&casa_token=oQo7yq1Y-EvkAAAAA:QgpQe6Ji25I99fkkTVotT3T6OvIKifjWZUFOTaeIuIbt5wpQuEojmJP hhpjI_IGyWdDyz85z

Canadian Department of National Defense and Armed Forces. (2022) *Code of values and ethics.* Available at www.canada.ca/en/department-national-defence/services/benefits-military/defence-ethics/policies-publications/code-value-ethics.html

Cantor, N. (1990) 'From thought to behaviour: 'having' and 'doing' in the study of personality and cognition', *American Psychologist,* 45, pp. 735–750.

Carr, D. (1999) 'Professional education and professional ethics', *Journal of Applied Philosophy,* 16 (1), pp. 33–46.

Carr, D. (2011) 'Virtue, Character and Emotion in People Professions: Towards a Virtue Ethics of Interpersonal Professional Conduct', in Bondi, L., Carr, D., Clark, C. and Clegg, C. (eds.) *Towards Professional Wisdom: Practical Deliberation in the People Professions,* Farnham: Ashgate.

Challans, T. (1999) 'Theory in practice: the possibility of a professional ethic'. Paper presented at the *Joint Services Conference on Professional Ethics,* 20 January. Available at http://isme.tamu.edu/JSCOPE99/Challans99.html.

Chapa, J.O. (2022) *Is Remote Warfare Moral?* New York: Hachette Book Group.

Claypool, G.A., Fetyko, D.F., and Pearson, M.A. (1990) 'Reactions to ethical dilemmas: a study pertaining to certified public accountants', *Journal of Business Ethics,* 9, pp. 699–706.

Coleman, S. (2015) 'Possible ethical problems with military use of non-lethal weapons', *Case Western Reserve Journal of International Law,* 47 (1), pp. 185–200.

Collings-Hughes, D., Townsend, D., and Williams, B. (2022) 'Professional codes of conduct: a scoping review', *Nursing Ethics,* 29 (1), pp. 19–34.

Collins, B. J. (2007) 'The officer corps and profession: time for a new model', *Joint Force Quarterly,* 45, pp. 104–110. Available at www.ndu.edu/press/jointForce-Quarterly.html

Cooke, S. and Carr, D. (2014) 'Virtue, practical wisdom and character in teaching', *British Journal of Educational Studies,* 62 (2), pp. 91–110.

Cox, R. (2018) 'The Ethics of War Up to Thomas Aquinas', in Lazar, S. and Frowe, H. (eds.), *The Oxford Handbook of Ethics of War.* Oxford: Oxford University Press, pp. 99–121.

Dandeker, C. and Freedman, L. (2002) 'The British Armed Services', *The Political Quarterly,* 73 (4), pp. 465–475.

DeFalco, J. and Doty, J. (2019) 'Considerations for Developing Self-Improving Systems to Support Phronesis: Moral and Ethical Thinking and Reasoning in Military Populations', in Sinatra A. et al. (eds.), *Designing Recommendations for Intelligent Tutoring Systems: Volume 7 – Self-Improving Systems.* Orlando: US Army Research Laboratory, pp. 169–175. Available at: https://aisconsortium.com/wp-content/uploads/Design-Recommendations-for-ITS_Volume-7-Self-Improving-Systems.pdf#page=169

Dixon-Woods, M., Yeung, K. and Bosk, C.L. (2011) 'Why is UK medicine No Longer a Self-Regulating Profession? The role of scandals involving "bad apple" doctors', *Social Science & Medicine*, 73 (10), pp. 1452–1459.

Ducich, S. (2018) 'Cyber force in an Ana[Law]G world: on self-defense, cyber operations, and the United States Law of war manual', *Homeland & National Security Law Review*, 6, pp. 21–58.

Duncan, J.C. (1999) 'The commander's role in developing rules of engagement', *Naval War College Review*, 52 (3), pp. 76–89. URL: www.jstor.org/stable/44643010

Dunlap, C.J. (2016) 'The DoD law of war manual and its critics: some observations', *International Law Studies*, 92 (85), pp. 85–118.

Eco, U. (1991) Reflections on War, La Rivista dei libri (1 April 1991); reprinted in U. Eco, Five Moral Pieces, trans. A. McEwen. New York: Harcourt, Inc., 1997, pp. 1–17.

Edmunds, T. (2006) 'What are armed forces for? The changing nature of military roles in Europe', *International Affairs*, 82 (6), pp. 1059–1075.

Edwards, E., Brantley, C., and Ruffin, P.B. (2017) 'Overview of Nanotechnology in Military and Aerospace Applications', in Mensah, T.O., Wang, B., Bothun, G., Winter, J., and Davis, V. (eds.), *Nanotechnology Commercialization: Manufacturing Processes and Products*. Hoboken: Wiley.

Emerton, P. and Handfield, T. (2018) 'Humanitarian Intervention and the Modern State System', in Lazar, S. and Frowe, H. (eds.), *The Oxford Handbook of Ethics of War*. New York: Oxford University Press, pp. 223–241.

Fish, D. and de Cossart, L. (2013) *Reflection for Medical Appraisal exploring and developing your clinical expertise and professional identity*. Aneumi Publications.

Fisher, D. (2011) *Morality and War: Can War be Just in the Twenty-first Century?* Oxford: Oxford University Press.

Flanagan, O. (1991) *Varieties of Moral Personalities*. Cambridge: Harvard University Press.

Fletcher, J.D. (2004) 'Cognitive Readiness: Preparing for the Unexpected', in Toiskallio, J. (ed), *Identity, Ethics, and Soldiership*. Helsinki: Finnish National Defence College.

Forster, A. (2006) 'Breaking the Covenant: Governance of the British Army in the Twenty-First Century', *International Affairs*, 82 (6), pp. 1043–2346.

Frankfurt, S. and Frazier, P. (2016, May 23) 'A review of research on moral injury in combat veterans', *Military Psychology*. Advance online publication. http://dx.doi.org/10.1037/mil0000132.

Frederick, B. and Johnson, D.E. (2015) *The Continued Evolution of U.S. Law of Armed Conflict Implementation Implications for the U.S. Military*. Santa Monica: Rand.

French, S.E. (2005) *The Code of the Warrior – Exploring Warrior Values Past and Present*. Lanham, New York: Rowman & Littlefield Publishers.

Frowe, H. (2011) *The Ethics of War and Peace: An Introduction*. London and New York: Routledge.

Frowe, H. (2018) 'The Just War Framework', in Lazar, S. and Frowe, H. (eds.), *The Oxford Handbook of Ethics of War*. New York: Oxford University Press, pp. 41–58.

Furlong, W., Crossan, M., Gandz, J. and Crossan, I. (2017) 'Character's essential role in addressing misconduct in financial institutions', *Bus. I. Int'l*. 18, pp. 199.

Gillies, J. (2005) 'Getting it right in the consultation: Hippocrates's problems, Aristotle's solution'. Occasional Paper 86. RGCP.

Gilman, S.C. (2005) 'Ethics codes and codes of conduct as tools for promoting an ethical and professional public service: comparative successes and lessons', *Talk given at the World Bank*, Washington DC.

Glazier, D., Colakovic, Z., Gonzalez, A., and Tripodes, Z. (2017) 'Failing our troops: critical assessment of the department of defense law of war manual', *Yale Journal of International Law*, 42 (2), pp. 215–278.

Goulding, V.J. (2000) 'Back to the Future with Asymmetric Warfare', *Parameters, Winter*, 30 (4), pp. 21–30.

Greenwood, E. (1957) 'Attributes of a profession', *Social Work*, 2 (3), pp. 45–55.

Griffen, B.J., Purcell, N., Burkman, K., Litz, B.T., Bryan, C.J., Schmitz, M., Villierme, C., Walsh, J., and Maguen, S. (2019) 'Moral injury: an integrative review', *Journal of Traumatic Stress*, 32, pp. 350–362. www.niwrc.org/sites/default/files/images/resource/moral_injury-_an_integrative_review_2019.pdf

Hackett, J.W. (1983) *The Profession of Arms*. New York: Macmillan.

Hajjar, R.M. (2014) 'Emergent postmodern US military culture', *Armed Forces & Society*, 40 (1), pp. 118–145.

Halloran, S.M. (1982) 'Aristotle's concept of ethos, or if not his, somebody else's', *Rhetoric Review*, 1 (1). Available at www.tandfonline.com/doi/abs/10.1080/073501 98209359037?journalCode=hrhr20

Hannah, S.T. and Avolio, B.J. (2011) 'Leader character, ethos, and virtue: individual and collective considerations', *The Leadership Quarterly*, 22 (5), pp. 989–994. Available at www.sciencedirect.com/science/article/pii/S1048984311001202?casa_token=3Bhv4-7KMm0AAAAA:Ck-cIiqOa4LnFIEM124_y-Bx_vHQzm3Ddqa-uGpk_Zc1gb0tJbYKjvUayAKNlzwvL7uHkXLM

Hannah, S.T., Campbell, D.J., and Matthews, M.D. (2010) 'Advancing a research agenda for leadership in dangerous contexts', *Military Psychology*, 22, pp. S157–S189. DOI:10.1080/08995601003644452.

Hannah, S.T. and Jennings, P.L. (2013) 'Leader ethos and big-C character', *Organizational Dynamics*, 42, pp. 8–16. DOI:dx.doi.org/10.1016/j.orgdyn.2012.12.002.

Headquarters Department of the [US] Army (2011) *The warrior ethos*. Available at: www.army.mil/article/50082/warrior_ethos#:~:text=The%20Army%20Warrior%20Ethos%20states,by%20which%20every%20Soldier%20lives.

Higgs-Kleyn, N. and Kapelianis, D. (1999) 'The role of professional codes in regulating ethical conduct', *Journal of Business Ethics*, 19 (4), pp. 363–374.

Holbeche, L. and Springett, N. (2004) '*In Search of Meaning at Work*', Roffey Park Institute, Horsham. http://citeseerx.ist.psu.edu/viewdoc/download?doi=10.1.1.458.1538&rep=rep1&type=pdf; Accessed 18th May 2018.

Horowitz, M.C. (2019) 'When speed kills: lethal autonomous weapon systems, deterrence and stability', *Journal of Strategic Studies*, 42 (6), pp. 764–788. DOI:10.1080/01402390.2019.1621174.

Huntington, S. (1957) *The Soldier and the State: The Theory and Practice of Civil-Military Relations*. Cambridge: Harvard University Press.

Huntington, S.P. (1957). *The Soldier and the State: The Theory and Practice of Civil-Military Relations*. Cambridge: Belknap Press. Pp 7.

Huntington, S.P. (1963) 'Power, expertise and the military profession', *Daedalus*, 92 (4), pp. 785–807.

Janowitz, M. (1960) *The Professional Soldier: A Social and Political Portrait*. New York: Free Press.

Jarrett, T. (2008) 'Warrior Resilience Training in Operation Iraqi Freedom: Combining Rational Emotive Behavior Therapy, Resiliency and Positive Psychology', *U.S. Army Medical Department Journal*, July/Sept. 2008, pp. 32–38.

Jauchem, J.R. and Cook, M.C. (2007) 'High-intensity acoustics for military nonlethal applications: a lack of useful systems', *Military Medicine*, 172 (2), pp. 182–189.

Jubilee Centre for Character and Virtues (2017) *A Framework for Character Education in Schools*. University of Birmingham, Jubilee Centre for Character and Virtues. https://www.jubileecentre.ac.uk/userfiles/jubileecentre/pdf/character-education/Framework%20for%20Character%20Education.pdf; accessed 14th August 2019.

Kaag, J. and Kreps, S. (2014) *Drone Warfare*. Cambridge, UK: Polity Press, pp. 11–17 [specific to technology p. 11]

Kaldor, M. (1999) *New and Old Wars: Organized Violence in a Global Era*. Cambridge: Polity Press.

Kateb, G. (2004) 'Courage as a Virtue', *Social Research*, 7 (1), pp. 39–72.

Kerr, S. (2021) Developing and testing a teaching intermediate concept measure of moral functioning: a preliminary reliability and validity study. *Ethics & Behavior*, 31 (5), pp. 350–364, DOI: 10.1080/10508422.2020.1794870.

King, P.M. and Mayhew, M.J. (2002) 'Moral judgement development in higher education: insights from the defining issues test', *Journal of Moral Education*, 31 (3), pp. 247–270.

Kinsella, E. A. and Pitman, A. (2012) 'Engaging phronesis in professional practice and education', in E. A. Kinsella and A. Pitman (eds.) *Phronesis as Professional Knowledge: Practical Wisdom in the Professions*. Rotterdam: Sense, pp. 1–13.

Kirkpatrick, J. (2015a) 'Drones and the Martial Virtue Courage', *Journal of Military Ethics*, 14 (3–4), pp. 202–219.

Kirkpatrick, J. (2015b) 'Reply to Sparrow: Martial Courage – or Merely Courage?', *Journal of Military Ethics*, 14 (3–4), pp. 228–231.

Kiszely, J. (2009) 'Postmodern challenges for modern warriors', *Army History*, 71 (1), pp. 19–33.

Kotzee, B., Paton, A., and Conroy, M. (2016) 'Towards an empirically informed account of phronesis in medicine', *Perspectives in Biology and Medicine*, 59 (3), pp. 337–350.

Krishnan, A. (2009) *Killer Robots: Legality and Ethicality of Autonomous Weapons*. London: Routledge.

Kristjánsson, K. (2015a) 'Phronesis as an ideal in professional medical ethics: some preliminary positionings and problematics', *Theoretical Medicine and Bioethics*, 36 (5), pp. 299–320.

Kristjánsson, K. (2015b) *Aristotelian Character Education*. Abingdon: Routledge.

Kristjánsson, K. (2017) *Aristotelian Character Education*. London: Routledge, pp. 14–15, 17, 28.

Kristjánsson, K., Arthur, J., Moller, F., and Huo, Y. (2017a) *Character and Virtues in Business and Finance*. Jubilee Centre for Character and Virtues.

Kristjánsson, K., Varghese, J., Arthur, J., and Moller, F. with Ferkany, M. (2017b) *Virtuous Practice in Nursing*. Jubilee Centre for Character and Virtues.

Lazar, S. (2017) 'War', in Zalta, E.N. (ed), *The Stanford Encyclopedia of Philosophy*. Available at https://plato.stanford.edu/archives/spr2017/entries/war/

Levine, S.D. and Rutigliano, J.A. (2015) 'U.S. military use of non-lethal weapons: reality vs perceptions', *Case Western Reserve Journal of International Law*, 47 (1), pp. 239–264.

Lewer, N. and Davison, N. (2005) 'Non-lethal technologies – an overview', *Disarmament Forum*, 1, pp. 37–51.

Litz, B.T., Stein, N., Delaney, E., Lebowitz, L, Nash, W.P., Silva, C., and Maguen, S. (2009) 'Moral injury and moral repair in war veterans: a preliminary model and intervention strategy', *Clinical Psychology Review*, 29 (8), pp. 695–706. www.sciencedirect.com/science/article/pii/S0272735809000920

LiVecche, M. (2021) *The Good Kill*. Oxford: Oxford University Press [xi – xii] and p. 25 for def. for suicide 35–37.

Lucas, G.R. (2010) 'Postmodern war', *Journal of Military Ethics*, 9 (4), pp. 289–298.

Manigart, P. (2005) 'Risks and recruitment in postmodern armed forces: the case of Belgium', *Armed Forces & Society*, 31 (4), pp. 559–582.

McKie, A., Baguley, F., Guthrie, C., Jackson, C., Kirkpatrick, P., Laing, A., O'Brien, S., Taylor, R. and Wimpenny, P. (2012) 'Exploring clinical wisdom in nursing education', *Nursing Ethics*, 19 (2), pp. 252–267.

Mechler, H.S. and Thoma, S.J. (2013) 'Moral Development Theory: Neo-Kohlbergian Theory', in Irby, B.J., Brown, G., Lara-Alecio, R., and Jackson, S. (eds.), *The Handbook of Educational Theories*. IAP Information Age Publishing, pp. 643–651.

Meier, M.W. (2016) 'Lethal Autonomous Weapons Systems (Laws): conducting comprehensive weapons review', *Temple International & Comparative Law Journal*, 30 (1), pp. 119–132.

Meredith, L.S., Sherbourne, C.D., Gaillot, S.J., Hansell, L., Ritschard, H.V., Parker, A.M. and Wrenn, G. (2011) *Promoting Psychological Resilience in the U.S. Military*. Santa Monica, CA: Centre for Military Health Policy Research.

Merriam-Webster (2022), 'Ethos', *Merriam-Webster.com Dictionary*. Available at: www.merriam-webster.com/dictionary/ethos

Micewski, E.R. (2005) 'Creativity and Military Leadership in Postmodern Times', Conference presentation at the *Conference for Interdisciplinary Creativity in Science*. In Bucharest, Hungary 25–26 February 2005.

Miller, J.J. (2004) 'Squaring the circle: teaching philosophical ethics in the military', *Journal of Military Ethics*, 3 (3), pp. 199–215. DOI:10.1080/15027570410006219.

Molestina, A., Ravichandran, K.R., and Welleck, M.N. (2020) 'Military applications of nanotechnology', *Student Papers in Public Policy*, 2 (1), pp. 1–22. Available at: https://docs.lib.purdue.edu/sppp/vol2/iss1/5

Mompeyssin, P. (2014) 'Soldiers' codes of conduct in different countries around the world. A comparative outlook', *Journal of Defense Resources Management*, 5 (1), pp. 5–9.

Moore, G. (2015) 'Corporate Character, Corporate Virtues', *Business Ethics: A European Review*, 24 (2), pp. 99–114.

Morgan, D.H.J. (1994) *Theater of War: Combat, the Military and Masculinities Theorising Masculinities*. B. H. K. M, London: Sage.

Moskos, C.C. (2000) 'Towards a Post-Modern Military?', in Cohen, S.A. (ed), *Democratic Societies and Their Armed Forces: Israel in Comparative Context*. London and Portland, OR: Frank Cass Publishers, pp. 3–26.

Moskos, C.C. and Burke, J. (1994) 'The Postmodern Military', in Burke, J. (ed), *The Military in New Times: Adapting Armed Forces to a Turbulent World*. London: Routledge, pp. 137–159.

Moskos, C.C. and Wood, F.R. (1988) *The Military More Than Just a Job?* Washington: Pergamon-Brassey's.

Moskos, C.C., Williams, J.A., and Segal, D.R. (1994) *The Postmodern Military: Armed Forces after the Cold War*. London and New York: Oxford University Press.

Moskos, C.C., Williams, J.A. and Segal, D.R. (2000) *The Postmodern Military: Armed Forces after the Cold War*, Oxford: Oxford University Press.

Moten, M. (2011) 'Who is a member of the military profession?', *Joint Force Quarterly*, 62 (3), pp. 14–17.

Narvaez, D. (2005) 'The Neo-Kohlbergian Tradition and Beyond: Schemas, Expertise, and Character', in Pope-Edwards, C. and Carlo, G. (eds.), *The Nebraska Symposium on Motivation, Vol. 51: Moral Motivation through Lifespan*. Lincoln, NE: University of Nebraska Press, pp. 119–164.

Nasu, H. and Faunce, T. (2009) 'Technology and the international law of weaponry: towards international regulation of nano-weapons', *Journal of Law, Information and Science*, 20, pp. 21–54.

Norwegian Armed Forces. (2022) *The values and standards of the Norwegian Armed Forces*. Available at www.forsvaret.no/en/about-us/missions-and values/values/Our%20Values%20and%20Standards.pdf/_/attachment/inline/2c0f78ef-ab8a-41c28aef93901ca107cf:67c7ed7ef5a9286d659518b6af56f4c5c42960c4/Our%20Values%20and%20Standards.pdf

Oakley, J. and Cocking, D. (2001) *Virtue Ethics and Professional Roles*. Cambridge: Cambridge University Press.

Oakley, J. and Cocking, D. (2002) *Virtue Ethics and Professional Roles*, Cambridge: Cambridge University Press.

Olson, L.W. (2014) *Towards a More Ethical Military: The Contribution of Aristotelian Virtue Theory to Military Ethics*, PhD, University of Texas at Austin.

Olsthoorn, P. (2005) 'Honour as a Motive for Making Sacrifices', *Journal of Military Ethics*, 3 (3), pp. 183–197.

Olsthoorn, P. (2007) 'Courage in the Military: Physical and Moral', *Journal of Military Ethics*, 6 (4), pp. 270–279.

Olsthoorn, P. (2011) *Military Ethics and Virtues: An Interdisciplinary Approach for the 21st Century*. London and New York: Routledge.

Orbons, S. (2012) 'Are non-lethal weapons a viable military option to strengthen the hearts and minds approach in Afghanistan?', *Defense and Security Analysis*, 28 (2) pp. 114–130.

Orend, B. (2002) *Human Rights: Concept and Context*. Peterborough, ON: Broadview Press.

Orend, B. (2013) *The Morality of War*. 2nd ed. Peterborough, ON: Broadview Press.

Osiel, M.J. (2002) *Obeying Orders: Atrocity, Military Discipline and the Law of War*. New Brunswick, New Jersey: Transaction.

Parsons, S. (2021) *Virtuous soldiers: is the current ethical training sufficient for the United States Army, or is a character development program what soldiers and officers need?*, Ph.D., University of Birmingham. Available at: https://etheses.bham.ac.uk/id/eprint/11481/.

Pellegrino, E. and Thomasma, D. (1993) *The Virtues in Medical Practice*. Oxford: Oxford University Press.

Peterson, C., & Seligman, M. E. P. (2004). *Character strengths and virtues: A handbook and classification*. Oxford University Press; American Psychological Association.

Pinijphon, P. (2009) *An ICM approach to the assessment of a medical ethics intervention in Thailand*, Ph.D., University of Alabama.

Pitman, A. (2012) 'Professionalism and professionalization: hostile ground for growing phronesis', in E. A. Kinsella and A. Pitman (eds.) *Phronesis as Professional Knowledge: Practical Wisdom in the Professions*. Rotterdam: Sense, pp. 131–146.

Pitschmann, V. and Hon, Z. (2016) 'Military importance of natural toxins and their analogs', *Molecules*, 21, pp. 1–20. DOI:10.3390/molecules21050556.

Ramsden, J. (2012) 'Nanotechnology for military applications', *Nanotechnology Perceptions*, 8 (2), pp. 99–131.

Ramsey, B. (2018) 'An ethical decision-making tool for offensive cyberspace operations', *Air and Space Power Journal*, 32 (3), pp. 62–71.

Rest, J.R., Narvaez, D., Bebeau, M., and Thoma, S.J. (1999a) 'A Neo-Kohlbergian approach: the DIT and schema theory', *Educational Psychology Review*, 11 (4), pp. 291–324.

Rest, J.R., Narvaez, D., Thoma, S.J., and Bebeau, M. (1999b) 'DIT2: devising and testing a revised instrument of moral judgment', *Journal of Educational Psychology*, 91 (4), pp. 644–659.

Rest, J.R., Narvaez, D., Thoma, S.J., and Bebeau, M.J. (2000) 'A Neo-Kohlbergian approach to morality research', *Journal of Moral Education*, 29 (4), pp. 381–395.

Rhodes, R. (2020) *The Trusted Doctor: Medical Ethics and Professionalism*. New York, NY: Oxford University Press.

Roach, J.A. (1983) 'Rules of engagement', *Naval War College Review*, 36 (1), pp. 46–55 (10 pages). www.jstor.org/stable/44642842

Robinson, P. (2007) 'Ethics training and development in the military', *Parameters: US Army War College*, 37 (1), pp. 23–36.

Robinson, P. (2008) 'Introduction: Ethics Education in the Military', in Robinson, P., de Lee, N., and Carrick, D. (eds.), *Ethics Education in the Military*. London: Ashgate, pp. 1–12.

Robinson, P. (2015) 'Determining the Limits of Moral Compromise: The Case of the Impunity of Afghanistan's Indigenous Security Forces', *Journal of Military Ethics*, 14 (3–4), pp. 276–279.

Roche, C., Thoma, S.J. and Wingfield, J. (2014) From workshop to e-learning: Using technology-enhanced 'Intermediate Concept Measure' as a framework for pharmacy ethics education and assessment. *Pharmacy*, 2014(2), 137–160.

Roff, H.M. (2014) 'The strategic robot problem: lethal autonomous weapons in war', *Journal of Military Ethics*, 13 (3), pp. 211–227. DOI:10.1080/15027570.2014.975 010.

Ross, L. and Nisbett, R. (1991) *The Person and the Situation: Perspectives of Social Psychology*. London: McGraw-Hill.

Royal Pharmaceutical Society (2011) *Reducing Workplace Pressure Through Professional Empowerment*. https://www.rpharms.com/Portals/0/RPS%20document%20library/Open%20access/Support/64585_Reducing%20Workplace%20Pressure%20through%20professional%20empowerment%20-%20FINAL. PDF?ver=2017-05-16-133220-000; accessed 20th November 2019.

Russell, D. (2009) *Practical Intelligence and the Virtues*. Oxford: Oxford University Press.

Sandin, P. (2007) 'Collective Military Virtues', *Journal of Military Ethics*, 6 (4), pp. 303–314.

Sayler, K.M. (2020) 'Defense primer: U.S. policy on lethal autonomous weapon systems', *Congressional Research Service (CRS)*, Washington DC: Library of Congress. Available at: https://crsreports.congress.gov

Schulzke, M. (2013) 'Ethically insoluble dilemmas in war', *Journal of Military Ethics*, 12 (2), pp. 95–110.

Schwartz, B. (2009) 'Our loss of wisdom', *TED2009*. https://www.ted.com/talks/barry_schwartz_our_loss_of_wisdom?; accessed 24th November 2019.

Schwartz, B. (2011) 'Using our practical wisdom', *TEDSalon New York*. https://www.ted.com/talks/barry_schwartz_using_our_practical_wisdom?language=en; accessed 24th November 2019.

Seijts, G., Crossan, M. and Carleton, E. (2017) 'Embedding leader character into HR practices and to achieve sustained excellence', *Organizational Dynamics*, 44 (1), pp. 65–74.

Seligman, M. (2011) Flourish – *A New Understanding of Happiness and Well-being - and How to Achieve Them*. London: Nicholas Brealey Publishing.

Sellman, D. (2009) 'Practical wisdom in health and social care: teaching for professional phronesis', *Learning in Health and Social Care*, 8 (2), pp. 84–91.

Sellman, D. (2012) 'Reclaiming competence for professional phronesis', in E. A. Kinsella and A. Pitman (eds.) *Phronesis as Professional Knowledge: Practical Wisdom in the Professions*. Rotterdam: Sense, pp. 115–130.

Shaw, M. (2005) *The New Western Way of War*. Cambridge: Polity Press.

Sherman, N. (2015) *Afterwar*. Oxford: Oxford: Oxford University Press.

Shields, P. (1991) Socio-economics: A Paradigm for Military Policy, in *Faculty Publications-Political Science*, Maryland, 11–13 October [Online], [Available at: http://ecommons.txstate.edu/polsfacp/11 [Accessed: 30 April 2022].

Shils, E. and Janowitz, M. (1975) 'Cohesion and Disintegration in the Wehrmacht in World War II', In Janowitz, M. (ed.) *Military Conflict: Essays in the Institutional Analysis of War and Peace*, London: Sage.

Snider, D.M. (2000) 'America's post modern military', *World Policy Journal*, 17 (1), pp. 47–54. Available at: www.jstor.org/stable/pdf/40209676.pdf

Snider, D.M. (2010) 'The Army Is not a Profession just because It Says It Is', in Wiggins, M.H. and Dabeck, L. (eds.), *Fort Leavenworth Ethics Symposium: Exploring the Professional Military Ethic*. Available at: https://d1wqtxts1xzle7.cloudfront.net/8655827/ethicsreport2010-with-cover-pagev2.pdf?Expires=1654857035&Signature=Ad-LIT~tc7sxYjCJo1lA7uIY8mpUFcFKSGUKFS4Da8vxZ1ZT3nqfVf0hFUd5GqOQ1QymRZaJVjNYIFHCINALQKdcLQ2h5r~tE5C2LOuNWWZUyCNSPmZel90LUOcM~Vyh1GMltgo6gn4qxhWYT0Tdks7sKyIBkM-tBJb3LMdjUJXOFiEWVuBvfxt4GqmQVPvH4PN~BqpXF6N83UQXburD4gRksQmOYxYqkbmND7WarM4Fn6RGkuinKbSuGu0g40hcqckL1VdHYk3g1JcbLle7kABkNbysdwKsylbrDHO8H7s-anOvnlTE-bMeJinTpmd6RSRvWSf6Qo644qDYhIcCZJeQ__&Key-Pair-Id=APKAJLOHF5GGSLRBV4ZA#page=37

Snider, D.M., Nagl, J.A., and Pfaff, C.A. (1999) *Army Professionalism, the Military Ethic, and Officership in the 21st Century*. Carlisle Barracks: United States Army

War College Press. Available at: https://press.armywarcollege.edu/cgi/viewcontent. cgi?article=1145&context=monographs

Snider, D.M. and Watkins, G.L. (2000) 'The future of army professionalism: a need for renewal and redefinition', *Parameters*, 30 (3), pp. 5–20. Available online at: https:// press.armywarcollege.edu/cgi/viewcontent.cgi?article=1996&context=parameters

Snider, D.M. and Watkins, G.L. (2002) 'Introduction', in Snider, D.M. and Matthews, L.J. (eds.), *The Future of the Army Profession*. Boston: McGraw-Hill Primis Custom Publishing.

Solis, G. (2010) 'Law of Armed Conflict's Four Core Principles', in *The Law of Armed Conflict: International Humanitarian Law in War*. Cambridge: Cambridge University Press, pp. 250–300. DOI:10.1017/CBO9780511757839.009.

Solis, G.D. (2014) 'Cyber warfare', *Military Law Review*, 219, pp. 1–52.

Sparrow, R. (2015) 'Martial and Moral Courage in Teleoperated Warfare: A Commentary on Kirkpatrick', *Journal of Military Ethics*, 14 (3–4), pp. 220–227.

Strachan, H. (2003) 'The Civil-Military Gap in Britain', *The Journal of Strategic Studies*, 26 (2), pp. 43–63.

Stratman, D. (2018) 'Drones and Robots', in Lazar, S. and Frowe, H. (eds.), *The Oxford Handbook of Ethics of War*. New York: Oxford University Press, pp. 473–487.

Surber, R. (2018) Artificial intelligence: autonomous technology (AT), lethal autonomous weapons systems (LAWS) and peace time threats. *ICT4Peace Foundation and the Zurich Hub for Ethics and Technology (ZHET)*, pp. 1–21. Available at: https:// ict4peace.org/wp-content/uploads/2018/02/2018_RSurber_AI-AT-LAWS-Peace-Time-Threats_final.pdf

Swain, R.M. and Pierce, A.C. (2019) 'The Profession of Arms', in Swain, R.M. and Pierce, A.C. (eds.), *The Armed Forces Officer: Essays on Leadership, Command, Oath, and Service Identity*. New York: Skyhorse Publishing, pp. 15–28.

Tate, J.S., Espinoza, S., Habbit, D., Hanks, C., Trybula, W., and Fazarro, D. (2015) 'Military and national security implications of nanotechnology', *Journal of Technology Studies*, 41 (1), pp. 20–28.

Thoma, S.J. (1986) 'Estimating gender differences in the comprehension and preference of moral issues', *Developmental Review*, 6, p. 165.

Thoma, S.J. (2014) 'Measuring moral thinking from a neo-Kohlbergian perspective', *Theory and Research in Education*, 12 (3), pp. 347–365.

Thoma, S.J., Derryberry, P., and Crowson, H.M. (2013) 'Describing and testing an intermediate concept measure of adolescent moral thinking', *European Journal of Developmental Psychology*, 10 (2), pp. 239–252.

Tripodi, P. (2012) 'Deconstructing the Evil Zone: How Ordinary Individuals Can Commit Atrocities', in Tripodi, P. and Wolfendale, J. (eds.) *New Wars and New Soldiers: Military Ethics in the Comtemporary World*, Surrey: Ashgate, pp. 201–216.

Tuckett, A.G. (2000) 'Virtuous principles as an ethic for nursing', *Contemp Nurse*, 9 (2), pp. 106–114. doi: 10.5172/conu.2000.9.2.106. PMID: 11854998.

Turner, M.E. (2008) *The development and testing of an army leader intermediate ethical concepts measure*, PhD, University of Alabama. Available at: https://pqdtopen. proquest.com/doc/304680070.html?FMT=ABS.

US Army. (2022) *Seven army values*. Available at: www.army.mil/values/

Verbruggen, M. (2019) 'The role of civilian innovation in the development of lethal autonomous weapon systems', *Global Policy*, 10 (3). DOI:10.1111/1758-5899.12663.

Verweij, D. (2007) 'Comrades or Friends? On Friendship in the Armed Forces', *Journal of Military Ethics*, 6 (4), pp. 280–291.

Walker, D.I. (2018) 'Character in the British Army: A Precarious Professional Practice', in Carr, D. (ed), *Cultivating Moral Character and Virtue in Professional Practice*. London and New York: Routledge, pp. 135–148.

Walker, D.I. (2020) 'Character and ethical judgment among junior army officers', *The Journal of Character and Leadership Development*, 7 (2), pp. 51–58.

Walker, D.I. (2021) 'Learning to own professional practice through character – the case of the junior British Army officer', *Journal of Moral Education*. DOI:10.1080/0305 7240.2021.1991292.

Walker, D.I., Thoma, S.J., and Arthur, J.A. (2021) 'Assessing ethical reasoning among junior British Army Officers using the Army Intermediate Concept Measure (AICM)', *Journal of Military Ethics*, 20 (1), pp. 2–20. DOI:10.1080/15027570.202 1.1895965.

Walker, D.I., Thoma, S.J., Jones, C., and Kristjánsson, K. (2017) 'Adolescent moral judgement: a study of UK secondary school pupils', *British Educational Research Journal*, 43 (3), pp. 588–607.

Walker, L.J. (2006) 'Gender and Morality', in Killen, M. and Smetana, J. G. (eds.), *Handbook of Moral Psychology*, pp. 93–118.

Walzer, M. (1977) *Just and Unjust Wars*. New York: Penguin.

Walzer, M. (2008) Interview with Michael Walzer. 'Michael Walzer on just War Theory'. Interview by *Big Think*, 15 April, Available at: https://bigthink.com/videos/michael-walzer-on-just-war-theory.

Watkins, J. (1999) 'Educating professionals: the changing role of UK professional associations', *Journal of Education and Work*, 12 (1), pp. 37–56.

West, A. (2017) 'The ethics of professional accountants: an Aristotelian perspective', *Accounting, Auditing & Accountability Journal*, 30 (2), pp. 328–351.

Wilensky, H.L. (1964) 'The professionalization of everyone?', *American Journal of Sociology*, 70 (2), pp. 137–158.

Williams, J.A. (2008) 'The military and society beyond the postmodern era', *Orbis*, 52 (2), pp. 5–9. DOI:10.1016/j.orbis.2008.01.003.

Wilson, I. and Meese, M.J. (2019) 'Officership and the Profession of Arms in the Twenty-First Century', in O'Connor, F.G., Schoomaker, E.B., and Smith, D.C. (eds.), *Fundamentals of Military Medicine*. Borden Institute, pp. 35–43.

Wolfendale, J. (2008) 'What is the Point of Teaching Ethics in the Military', in Robinson, P., De Lee, N. and Carrick, D. (eds.) *Ethics Education in the Military*, Aldershot: Ashgate, pp. 161–174.

Wolfendale, J. (2009) 'Professional Integrity and Disobedience in the Military', *Journal of Military Ethics*, 8 (2), pp. 127–140.

Worth, J. and Van Den Brande, J. (2019) *Teacher Labour Market in England: Annual Report 2019*. https://www.nfer.ac.uk/media/3344/teacher_labour_market_in_england_2019.pdf; accessed 20th November 2019.

Zacher, H., McKenna, B., Rooney, D., and Gold, S. (2015) 'Wisdom in the military context', *Military Psychology*, 27 (3), pp. 142–154. DOI:10.1037/mil0000070.

Zavaliy, A.G. and Aristidou, M. (2014) 'Courage: A Modern Look at an Ancient Virtue', *Journal of Military Ethics*, 13 (2), pp. 174–189.

Index

Note: Page numbers in *italics* indicate a figure and page numbers in **bold** indicate a table on the corresponding page.